What has been said about this book

" There aren't sufficient superlatives available to express how much this text helped my understanding. "

Student from San Diego, California

" My questions were easily answered by referring to this highly readable book. "

Calculus Student from Singapore

" Elegant in its presentation and demystification of the subject. Highly recommended to anyone beginning the study of calculus. "

Student from Qatar

" Read this book, and champion math fundamentals. "

Student from Seattle, Washington

" Lucid and unpretentious, incredible book. "

Student from North America

" This is a classic text. "

Student from Tennessee

" This book has helped me to learn so much already that I'm sure I can get an easy "A" in all of my math classes. "

Student from West Linn, Oregon

(More on next page)

What has been said about this book

(Continued from previous page)

" I flunked calculus twice, then got an "A" with this book!"
Finance Student from Chicago

" I am very happy to have found this book."
Calculus Student from New Jersey

" One of those rare math books that read like a thriller."
Student

" Wonderful! This is an excellent book."
Student from Plainview, Texas

" Should be about 20 stars! This is a great book!"
Student from Washington State

" This book encouraged me to pursue further and more challenging courses."
Student from Clinton, Iowa

" This book is the beginners ants pants of Calculus!"
Student from Australia

" I found the method and clarity of presentation to be refreshing, practical, and easy to learn."
A Calculus Reader

(More on back two pages)

CALCULUS

Silvanus P. Thompson, F.R.S.
Max Fogiel, Ph.D.

NEW EDITION makes
*Calculus much easier
to learn!*

Research & Education Association
61 Ethel Road West
Piscataway, New Jersey 08854

SUPER TEXTBOOK™
OF CALCULUS

Printed in the United States of America

Library of Congress Control Number 2002101792

International Standard Book Number 0-87891-781-0

SUPER TEXTBOOK is a trademark of
Research & Education Association, Piscataway, New Jersey 08854

WHAT THIS Super Textbook™ WILL DO FOR YOU

- This book is designed for students who take only the first year of calculus. Hence, it is less bulky and overwhelming than the typical calculus textbooks.

- The book is the first of its kind to answer questions that students have long been wondering about. For example, students often ask why they need to study calculus in the first place. This is answered in the introductory section "Why We Study Calculus". The explanation is given in a manner not found in other textbooks. The usual textbooks avoid this all-important question altogether or the explanations given are so abstract and complicated that they cannot be understood by students embarking on a first-year study of calculus.

- Another new feature, not found in the usual textbooks, is that the authors have kept in mind that a student cannot be expected to have previous familiarity with any of the topics covered. To provide for better understanding, separate new sections are included throughout the book on "Why We Study This Topic" and "How This Topic is Related to Preceding Ones."

- The book builds on the popular and classic *Thompson* edition, which has a long history of proven success with first-year calculus students.

- This newly revised edition combines the proven teaching/learning techniques of the past with new approaches and explanations that have been shown to work well.

- The book is written with a "light touch" to make it user friendly. It avoids the abstract and hard-to-follow explanations found in the usual calculus texts. It can help students relax about calculus. It even makes the subject interesting and challenging.

- Numerous applications with practical examples are included to help the student understand each topic quickly and thoroughly.

- The book also includes a large number of problems with solutions in step-by-step detail to help the student with homework assignments and in preparing for exams.

Dr. Max Fogiel

TABLE OF CONTENTS

Chapter	Page

iv

vi

INTRODUCTION

Why We Study Calculus

Calculus is a powerful tool through which we can determine precisely the behavior of variables.

For example, suppose we need to find the area of a floor rug that has a varying perimeter as shown in Fig. 1.

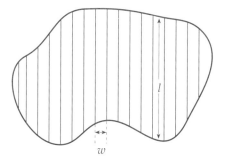

l

w

FIG. 1

One way to find the average approximate area is to divide the area into a series of rectangles of width w and length l. As may be seen from Figure 1, that if the widths of such rectangles are made the same, then the lengths of the rectangles vary in accordance with the variation in the rug's perimeter. The area of each such rectangle is wl. If we add up the areas of all the rectangles, then we obtain an approximate total area for the rug. As may be seen, the ends of every rectangle are not perpendicular to the sides l. Instead the ends slope with respect to the sides l in accordance with the slopes of the perimeter. Consequently, the area of each rectangle is not truly $w \times l$. As a result, the area of each rectangle taken as wl, is only an approximation and therefore the total area is also only an approximation. The approximation can be refined by making the width narrower. As long as the width w is some finite amount, however, the total area remains at best an approximation.

By using calculus techniques it is possible to obtain the total area precisely, and it is not necessary to add up laboriously all of the areas of the individual rectangles.

Through the use of calculus it is possible to make the width w infinitesimally small and therefore the amount of rectangles substantially infinite in number. As a result, the total area can be determined precisely.

As another example, assume that it is necessary to determine the volume of an object that has a curved surface such as a pear, shown in Fig. 2.

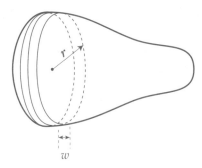

FIG. 2

One way of determining the volume of the pear is to divide the pear into disk-shaped slices having a radius r and a width w. The volume of each slice is the area, πr^2, of the slice multiplied by the thickness w. By adding up all of the volumes of the individual slices, we can arrive at an approximate total volume for the pear. The reason for being only approximate is that the edges of the slices are curved instead of being straight-lined and perpendicular to the disk-shaped faces of the slices.

Again the approximation can be refined by making the slices thinner. However, as long as the thickness of the slice is some finite amount, the total volume remains at best an approximation.

With calculus techniques, on the other hand, it is possible to obtain the total volume precisely by making the slices infinitesimally thin. By using calculus it is not necessary to add up the volumes of a large number of individual slices to result in a desired approximation – not a precise amount.

In a still further example, consider a racing car driving along a track at varying speed. The racing car is clocked at uniform intervals along the track, and a resultant graph of the position of the car versus time is given in Fig. 3.

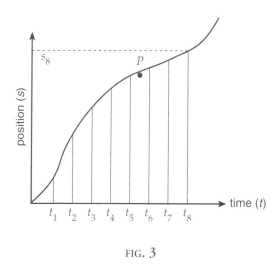

FIG. 3

If one were to divide the position s_8 by the time t_8, one could obtain a speed $\dfrac{s_8}{t_8}$ which one might interpret to be the car's speed in traveling from the starting position at time t_0 to time t_8. Such a speed calculation, however, doesn't tell us anything about the speed of the car at any point along its path from s_0 to s_8. In Figure 3, the speed of the car varied throughout its path. If we were to want the speed between time instants t_4 and t_5, for example, we could compute the ratio $\dfrac{s_5 - s_4}{t_5 - t_4}$ to obtain an average speed during this interval. It can be expected that this average speed may differ from the speed computed from $\dfrac{s_8}{t_8}$.

The ratio $\dfrac{s_5 - s_4}{t_5 - t_4}$, however, does not reveal the car's speed at a specific instant between the interval $t_5 - t_4$, such as at point P. So far we have assumed that the speed is uniform during this interval, implying that the graph in Figure 3 is a straight line between $t_5 - t_4$. As seen in Figure 3, however, the graph is curved, instead, within the interval $t_5 - t_4$. The ratio $\dfrac{s_5 - s_4}{t_5 - t_4}$, therefore, does not give accurately the speed at point P. The accuracy can be improved, however, by making the interval $t_5 - t_4$ small. The smaller this interval, the greater is the accuracy for the computed speed at point P.

To obtain the precise speed at point P, we can use calculus through which we allow the time interval $t_5 - t_4$ to approach zero and become infinitesimally small. Using calculus, therefore, we can obtain a precise result here also.

By using calculus we can study the behavior of functions with varying characteristics and obtain precise results with relative ease. Such functions can be encountered in science and engineering, including physics, chemistry, as well as in economics, biology, psychology, and the social sciences. By using calculus, moreover, it is not necessary to deal with small increments and to laboriously sum up the many increments, for example. Calculus provides a technique through which precise results are obtainable in a rather quick manner which, however, may not always be effortless.

PROLOGUE

Considering how many fools can calculate, it is surprising that it should be thought either a difficult or a tedious task for any other fool to learn how to master the same tricks.

Some calculus-tricks are quite easy. Some are enormously difficult. The fools who write the text-books of advanced mathematics—and they are mostly clever fools—seldom take the trouble to show you how easy the easy calculations are. On the contrary, they seem to desire to impress you with their tremendous cleverness by going about it in the most difficult way.

Being myself a remarkably stupid fellow, I have had to unteach myself the difficulties, and now beg to present to my fellow fools the parts that are not hard. Master these thoroughly, and the rest will follow. What one fool can do, another can.

COMMON GREEK LETTERS USED AS SYMBOLS

Capital	Small	English Name	Capital	Small	English Name
A	α	Alpha	Λ	λ	Lambda
B	β	Beta	M	μ	Mu
Γ	γ	Gamma	Ξ	ξ	Xi
Δ	δ	Delta	Π	π	Pi
E	ϵ	Epsilon	P	ρ	Rho
H	η	Eta	Σ	σ	Sigma
Θ	θ	Thēta	Φ	ϕ	Phi
K	κ	Kappa	Ω	ω	Omega

Why We Study Functions

Functions are relationships/equations that appear throughout calculus. We work with functions in almost every problem in calculus involving limits, derivatives, and integrals. To understand calculus it is necessary to first understand functions. A function is generally denoted by $y = f(x)$ and pronounced as "y is a function of x". $f(x)$ can represent a wide range of equations. In any equation different values of x (the independent variable) will usually result in different values for y (the dependent variable).

1

CHAPTER 1

To Deliver You from the Preliminary Terrors

The preliminary terror, which chokes off most students from even attempting to learn how to calculate, can be abolished once for all by simply stating what is the meaning—in common-sense terms—of the two principal symbols that are used in calculating.

These dreadful symbols are:

(1) d which merely means "a little bit of".

Thus dx means a little bit of x; or du means a little bit of u. Ordinary mathematicians think it more polite to say "an element of", instead of "a little bit of". Just as you please. But you will find that these little bits (or elements) may be considered to be infinitesimal small.

(2) $\int$ which is merely a long S, and may be called (if you like) "the sum of".

Thus $\int dx$ means the sum of all the little bits of x; or $\int dt$ means the sum of all the little bits of t. Ordinary mathematicians call this symbol "the integral of". Now any fool can see that if x is considered as made up of a lot of little bits, each of which is called dx, if you add them all up together you get the sum of all the dx's (which is the same thing as the whole of x). The word "integral" simply means "the whole". If you think of the duration of time for one hour, you may (if you like) think of it as cut up into 3600 little bits called seconds.

The whole of the 3600 little bits added up together make one hour.

When you see an expression that begins with this terrifying symbol, you will henceforth know that it is put there merely to give you instructions that you are now to perform the operation (if you can) of totalling up all the little bits that are indicated by the symbols that follow.

That's all.

CHAPTER 2

On Different Degrees of Smallness

Why We Study Smallness/Limits and Its Relation to Differentiation and Integration

In the description that follows on "smallness," a little bit of x is referred to as dx. This symbol dx is not meant to be "d" multiplied by "x". Instead, it is intended to mean an increment of x which is often written as Δx.

In the introductory section "Why We Study Calculus," we talked about increments of time and incremental slices, for example. We mentioned that when using calculus techniques such increments become infinitesimal small to the extent that they approach zero in size.

The study of "limits" deals with situations when increments of a variable approach zero in size. In mathematical terms, this is often expressed, for example by

$$\frac{dy}{dx} = \lim_{\Delta x \to 0} \frac{f(x + \Delta x) - f(x)}{\Delta x}$$

The topic of limits is the most abstract one in the study of first-year calculus, which deals with the three parts: limits, differentiation, and integration. It is possible to perform the last two parts without a thorough knowledge of limits, and for this reason you may ask why are we learning this abstract topic of limits? The answer to that is that a knowledge of limits helps in a thorough understanding of calculus.

4

We shall find that in our processes of calculation we have to deal with small quantities of various degrees of smallness.

We shall have also to learn under what circumstances we may consider small quantities to be so minute that we may omit them from consideration. Everything depends upon relative minuteness.

Before we fix any rules let us think of some familiar cases. There are 60 minutes in the hour, 24 hours in the day, 7 days in the week. There are therefore 1440 minutes in the day and 10,080 minutes in the week.

Obviously 1 minute is a very small quantity of time compared with a whole week. Indeed, our forefathers considered it small as compared with an hour, and called it "one minute", meaning a minute fraction—namely one sixtieth—of an hour. When they came to require still smaller subdivisions of time, they divided each minute into 60 still smaller parts, which, in Queen Elizabeth's days, they called "second minutes" (*i.e.* small quantities of the second order of minuteness). Nowadays we call these small quantities of the second order of smallness "seconds". But few people know *why* they are so called.

Now if one minute is so small as compared with a whole day, how much smaller by comparison is one second!

Again, think of a dollar compared with a penny: it is worth only a part. A penny is of precious little importance compared with a dollar: $\frac{1}{100}$ it may certainly be regarded as a small quantity. But compare a penny with ten thousand dollars: relative to this greater sum, a penny is of no more importance than a hundredth of a penny would be to a hundred dollars. Even a hundred dollars is relatively a negligible quantity in the wealth of a millionaire.

Now if we fix upon any numerical fraction as constituting the proportion which for any purpose we call relatively small, we can easily state other fractions of a higher degree of smallness. Thus if, for the purpose of time, $\frac{1}{60}$ be called a *small* fraction, then $\frac{1}{60}$ of $\frac{1}{60}$ (being a *small* fraction of a *small* fraction) may be regarded as a *small quantity of the second order* of smallness.*

*The mathematicians may talk about the second order of "magnitude" (*i.e.* greatness) when they really mean second order of *smallness*. This is very confusing to beginners.

5

Or, if for any purpose we were to take 1 percent $\left(i.e.\ \tfrac{1}{100}\right)$ as a *small* fraction then 1 percent of 1 percent $\left(i.e.\ \tfrac{1}{10,000}\right)$ would be a small fraction of the second order of smallness; and $\tfrac{1}{1,000,000}$ would be a small fraction of the third order of smallness, being 1 percent of 1 percent of 1 percent.

Lastly, suppose that for some very precise purpose we should regard $\tfrac{1}{1,000,000}$ as "small". Thus, if a first-rate chronometer is not to lose or gain more than half a minute in a year, it must keep time with an accuracy of 1 part in 1,051,200. Now if, for such a purpose, we regard $\tfrac{1}{1,000,000}$ (or one millionth) as a small quantity, then $\tfrac{1}{1,000,000}$ of $\tfrac{1}{1,000,000}$, that is, $\tfrac{1}{1,000,000,000,000}$ will be a small quantity of the second order of smallness, and may be utterly disregarded, by comparison.

Then we see that the smaller a small quantity itself is, the more negligible does the corresponding small quantity of the second order become. Hence we know that *in all cases we are justified in neglecting the small quantities of the second—or third* (or higher)—*orders,* if only we take the small quantity of the first order small enough in itself.

But it must be remembered that small quantities, if they occur in our expressions as factors multiplied by some other factor, may become important if the other factor is itself large. Even a penny becomes important if only it is multiplied by a few hundred.

Now in the calculus we write dx for a little bit of x. These things such as dx, and du, and dy, are called "differentials", the differential of x, or of u, or of y, as the case may be. [You *read* them as *dee-eks*, or *dee-you*, or *dee-wy*.] If dx be a small bit of x, and relatively small of itself, it does not follow that such quantities as $x \cdot dx$, or $x^2 dx$, or $a^x dx$ are negligible. But $dx \times dx$ would be negligible, being a small quantity of the second order.

A very simple example will serve as illustration.

Let us think of x as a quantity that can grow by a small amount so as to become $x + dx$, where dx is the small increment added by growth. The square of this is $x^2 + 2x \cdot dx + (dx)^2$. The second term is not negligible because it is a first-order quantity; while the third term is of the second order of smallness, being a bit of a bit of x . Thus if we took dx to mean numerically, say, $\tfrac{1}{60}$ of x, then the second term would be $\tfrac{2}{60}$ of x^2, whereas the third term would be $\tfrac{1}{3,600}$ of x^2. This last term is clearly less important than the second. But if we go further and take dx to mean only $\frac{1}{1000}$ of x, then the second term will be $\frac{2}{1000}$ of x^2, while the third term will be only $\frac{1}{1,000,000}$ of x^2.

Geometrically this may be depicted as follows: Draw a square (Fig. 1) the side of which we will take to represent x. Now suppose the square to grow by having a bit dx added to its size each way. The enlarged square is made up of the original square x^2, the two rectangles at the top and on the right, each of which is of area $x \cdot dx$ (or together $2x \cdot dx$), and a little square at the top right-hand corner which is $(dx)^2$. In Fig. 2 we have taken dx as quite a big fraction of x—about $\frac{1}{5}$. But suppose we had taken it only $\frac{1}{100}$—about the thickness of an inked line drawn with a fine pen as in figure 3. Then the little corner square will have an area of only $\frac{1}{10,000}$ of x^2, and be practically invisible. Clearly $(dx)^2$ is negligible if only we consider the increment dx to be itself small enough.

Let us consider a simile.

Suppose a millionaire were to say to his secretary: next week I will give you a small fraction of any money that comes in to me. Suppose that the secretary were to say to his boy: I will give you a small fraction of what I get. Suppose the fraction in each case to be $\frac{1}{100}$ part. Now if Mr. Millionaire received during the next week $1,000, the secretary would receive $10 and the boy 1 dime. Ten dollars would be a small quantity com-

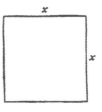

FIG. 1.

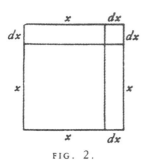

FIG. 2.

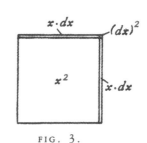

FIG. 3.

7

pared with $1,000; but a dime is a small small quantity indeed, of a very secondary order. But what would be the disproportion if the fraction, instead of being $\frac{1}{100}$, had been settled at $\frac{1}{1000}$ part? Then, while Mr. Millionaire got his $1,000, Mr. Secretary would get only $1.00, and the boy only a tenth of a penny!

The witty Dean Swift once wrote:

"So, Nat'ralists observe, a Flea
Hath smaller Pleas that on him prey.
And these have smaller Fleas to bite 'em.
And so proceed ad infinitum."

An ox might worry about a flea of ordinary size—a small creature of the first order of smallness. But he would probably not trouble himself about a flea's flea; being of the second order of smallness, it would be negligible. Even a gross of fleas' fleas would not be of much account to the ox.

Definition of Limits

Let f be a function that is defined on an open interval containing a, but possibly not defined at a itself. Let L be a real number. The statement

$$\lim_{x \to a} f(x) = L$$

defines the limit of the function f(x) at the point a. Very simply, L is the value that the function has as the point a is approached.

Additional Problem Solving Examples

Q Find $\lim_{x \to 2} f(x) = 2x + 1$

A As $x \to 2$, f(x) $\to$ 5. Therefore, $\lim_{x \to 2} (2x + 1) = 5$

Q Find $\lim_{x \to 3} f(x) = \dfrac{x^2 - 9}{x + 1}$

A $\lim_{x \to 3} \dfrac{x^2 - 9}{x + 1} = \dfrac{0}{4} = 0.$

8

Find $\lim_{x \to 0} (x\sqrt{x-3})$.

 In checking the function by simple substitution, we see that:

$$x\sqrt{x-3} = 0$$

if $\qquad x = 0$.

However, this function does not have real values for values of x less than 3. Therefore, since x cannot approach 0, f(x) does not approach 0 and the limit does not exist. This example illustrates that we cannot properly find

$$\lim_{x \to a} f(x)$$

by finding f(a), even though they are equal in many cases. We must consider values of x near a, but not equal to a.

9

CHAPTER 3

On Relative Growings

Why We Study Derivatives

You may recall from the introductory section, "Why We Study Calculus," that we can determine precisely the speed of a vehicle through the use of calculus techniques. The speed calculation is obtained by using derivatives. Speed of a vehicle, for example is equal to a change in distance traveled Δs over a corresponding change in time Δt, so that speed $= \dfrac{\Delta s}{\Delta t}$, as shown in the following graph. The symbol Δ is used here to denote increment or interval.

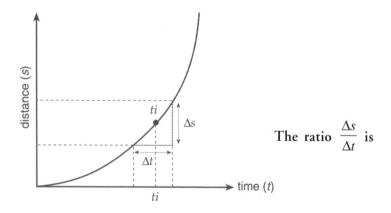

The ratio $\dfrac{\Delta s}{\Delta t}$ is

only an average of the speed that varies within the interval Δt. It is not a precise indication of the speed at any instant within the interval. We can obtain a more precise speed, however, by making the interval Δt small. The small interval Δt includes the time ti which is of interest to us.

By making Δt smaller and smaller, Δt finally approaches zero. Under this condition, we can obtain the precise speed at the time instant ti. At that point the speed is equal to the slope of the tangent at that point. The slope of the tangent is denoted by $\dfrac{ds}{dt}$, the derivative.

In general, derivatives denote rates or changes in a variable. In the preceding example, the rate is speed. Rates also appear in problems involving water flow, acceleration, economic changes, biological and physical changes, to mention a few.

With derivatives it is not necessary to go through the procedure of laboriously selecting small intervals, such as Δs and Δt, to obtain the rate or slope of the tangent. With the use of derivatives we can obtain rates or slopes of functions with relative ease.

All through the calculus we are dealing with quantities that are growing, and with rates of growth. We classify all quantities into two classes: *constants* and *variables*. Those which we regard as of fixed value, and call *constants*, we generally denote algebraically by letters from the beginning of the alphabet, such as *a*, *b*, or *c*; while those which we consider as capable of growing, or (as mathematicians say) of "varying", we denote by letters from the end of the alphabet, such as *x*, *y*, *z*, *u*, *v*, *w*, or sometimes *t*.

Moreover, we are usually dealing with more than one variable at once, and thinking of the way in which one variable depends on the other: for instance, we think of the way in which the height reached by a projectile depends on the time of attaining that height. Or, we are asked to consider a rectangle of given area, and to enquire how any increase in the length of it will compel a corresponding decrease in the breadth of it. Or, we think of the way in which any variation in the slope of a ladder will cause the height that it reaches, to vary.

Suppose we have got two such variables that depend on one another. An alteration in one will bring about an alteration in the other, *because* of this dependence. Let us call one of the variables *x*, and the other that depends on it *y*.

11

Suppose we make x to vary, that is to say, we either alter it or imagine it to be altered, by adding to it a bit which we call dx. We are thus causing x to become $x + dx$. Then, because x has been altered, y will have altered also, and will have become $y + dy$. Here the bit dy may be in some cases positive, in others negative; and it won't (except very rarely) be the same size as dx.

Take Two Examples.

(1) Let x and y be respectively the base and the height of a right-angled triangle (Fig. 4), of which the slope of the other side is fixed at 30°. If we suppose this triangle to expand and yet keep its angles the same as at first, then, when the base grows so as to become $x + dx$, the height be-

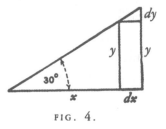

FIG. 4.

comes $y + dy$. Here, increasing x results in an increase of y. The little triangle, the height of which is dy, and the base of which is dx, is similar to the original triangle; and it is obvious that the value of the ratio $\dfrac{dy}{dx}$ is the same as that of the ratio $\dfrac{y}{x}$. As the angle is 30° it will be seen that here[1]

$$\frac{dy}{dx} = \frac{1}{1.73\ldots} = tan\ 30°$$

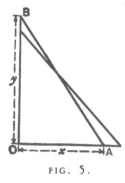

FIG. 5.

(2) Let x represent, in Fig. 5, the horizontal distance, from a wall, of the bottom end of a ladder, AB, of fixed length; and let y be the height it reaches up the wall. Now y clearly depends on x. It is easy to see that, if we pull the bottom end A a bit farther from the wall, the top end B will come down a little lower. Let us state this in scientific language. If we increase x to $x + dx$, then y will become $y - dy$; that is, when x receives a positive increment, the increment which results to v is negative.

12

Yes, but how much? Suppose the ladder was so long that when the bottom end A was 19 inches from the wall the top end B reached just 15 feet from the ground. Now, if you were to pull the bottom end out 1 inch more, how much would the top end come down? Put it all into inches: $x = 19$ inches, $y = 180$ inches. Now the increment of x which we call dx, is 1 inch: or $x + dx = 20$ inches.

How much will y be diminished? The new height will be $y - dy$. If we work out the height by the Pythagorean Theorem, then we shall be able to find how much dy will be. The length of the ladder is

$$\sqrt{(180)^2 + (19)^2} = 181 \text{ inches.}$$

Clearly, then, the new height, which is $y - dy$, will be such that

$$(y - dy)^2 = (181)^2 - (20)^2 = 32761 - 400 = 32361,$$
$$y - dy = \sqrt{32361} = 179.89 \text{ inches.}$$

Now y is 180, so that dy is $180 - 179.89 = 0.11$ inch.

So we see that making dx an increase of 1 inch has resulted in making dy a decrease of 0.11 inch.

And the ratio of dy to dx may be stated thus:

$$\frac{dy}{dx} = \frac{0.11}{1}$$

It is also easy to see that (except in one particular position) dy will be of a different size from dx.

Now right through the differential calculus we are hunting, hunting, hunting for a curious thing, a mere ratio, namely, the proportion which dy bears to dx when both of them are infinitesimal small.

It should be noted here that we can only find this ratio $\frac{dy}{dx}$ when y and x are related to each other in some way, so that whenever x varies y does vary also. For instance, in the first example just taken, if the base x of the triangle be made longer, the height y of the triangle becomes greater also, and in the second example, if the distance x of the foot of the ladder from the wall be made

13

to increase, the height y reached by the ladder decreases in a corresponding manner, slowly at first, but more and more rapidly as x becomes greater. In these cases the relation between x and y is perfectly definite, it can be expressed mathematically, being $\dfrac{y}{x} = \tan 30°$ and $x^2 + y^2 = l^2$ (where l is the length of the ladder) respectively, and $\dfrac{dy}{dx}$ has the meaning we found in each case.

If, while x is, as before, the distance of the foot of the ladder from the wall, y is, instead of the height reached, the horizontal length of the wall, or the number of bricks in it, or the number of years since it was built, any change in x would naturally cause no change whatever in y; in this case $\dfrac{dy}{dx}$ has no meaning whatever, and it is not possible to find an expression for it. Whenever we use differentials dx, dy, dz, etc., the existence of some kind of relation between x, y, z, etc., is implied, and this relation is called a "function" in x, y, z, etc.; the two expressions given above, for instance, namely, $\dfrac{y}{x} = \tan 30°$ and $x^2 + y^2 = l^2$, are functions of x and y. Such expressions contain implicitly (that is, contain without distinctly showing it) the means of expressing either x in terms of y or y in terms of x, and for this reason they are called *implicit functions in x and y*; they can be respectively put into the forms

$$y = x \tan 30° \quad \text{or} \quad x = \frac{y}{\tan 30°}$$

and $\qquad y = \sqrt{l^2 - x^2} \quad \text{or} \quad x = \sqrt{l^2 - y^2}.$

These last expressions state explicitly the value of x in terms of y, or of y in terms of x, and they are for this reason called *explicit functions* of x or y. For example $x^2 + 3 = 2y - 7$ is an implicit function of x and y; it may be written $y = \dfrac{x^2 + 10}{2}$ (explicit function of x) or $x = \sqrt{2y - 10}$ (explicit function of y). We see that an explicit function in x, y, z, etc., is simply something the value of which changes when x, y, z, etc., are changing, either one at the time or several together. Because of this, the value of the explicit function is called the *dependent variable,* as it depends on the value of the other variable quantities in the function; these other variables are called the *independent variables* because their value is not determined from the value assumed by the function. For example, if $u = x^2 \sin \theta$, x and θ are the independent variables, and u is the dependent variable.

Sometimes the exact relation between several quantities x, y, z either is not known or it is not convenient to state it; it is only known, or convenient to state, that there is some sort of relation between these variables, so that one cannot alter either x or y or z singly without affecting the other quantities; the existence of a function in x, y, z is then indicated by the notation $F(x, y, z)$ (implicit function) or by $x = F(y, z)$, $y = F(x, z)$ or $z = F(x, y)$ (explicit function). Sometimes the letter f or ϕ is used instead of F, so that $y = F(x)$, $y = f(x)$ and $y = \phi(x)$ all mean the same thing, namely, that the value of y depends on the value of x in some way which is not stated.

We call the ratio $\dfrac{dy}{dx}$, "the *differential coefficient* of y with respect to x". It is a solemn scientific name for this very simple thing. But we are not going to be frightened by solemn names, when the things themselves are so easy. Instead of being frightened we will simply pronounce a brief curse on the stupidity of giving long crack-jaw names; and, having relieved our minds, will go on to the simple thing itself, namely the ratio $\dfrac{dy}{dx}$.

In ordinary algebra which you learned at school, you were always hunting some unknown quantity which you called x or y; or sometimes there were two unknown quantities to be hunted for simultaneously. You have now to learn to go hunting in a new way; the fox being now neither x nor y. Instead of this you have to hunt for this curious cub called $\dfrac{dy}{dx}$. The process of finding the value of $\dfrac{dy}{dx}$ is called "differentiating". But, remember, what is wanted is the value of this ratio when both dy and dx are them-

15

selves infinitesimal small. The true value of the derivative is that to which it approximates in the limiting case when each of them is considered as infinitesimally minute.

Let us now learn how to go in quest of $\dfrac{dy}{dx}$.

How to Read Derivatives

It will never do to fall into the schoolboy error of thinking that dx means d times x, for d is not a factor—it means "an element of" or "a bit of" whatever follows. One reads dx thus: "dee-eks".

In case the reader has no one to guide him in such matters it may here be simply said that one reads derivatives in the following way. The derivative

$\dfrac{dy}{dx}$ is read *"dee-wy dee-eks"*, or *"dee-wy over dee-eks"*.

So also $\dfrac{du}{dt}$ is read *"dee-you dee-tee"*.

Second derivatives will be met with later on. They are like this: $\dfrac{d^2y}{dx^2}$; which is read *"dee-two-wy over dee-eks-squared"*, and it means that the operation of differentiating y with respect to x has been (or has to be) performed twice over.

Another way of indicating that a function has been differentiated is by placing an accent to the symbol of the function. Thus if $y = F(x)$, which means that y is some unspecified function of x, we may write $F'(x)$ instead of $\dfrac{d(F(x))}{dx}$. Similarly, $F''(x)$ will mean that the original function $F(x)$ has been differentiated twice over with respect to x.

The Derivative and Δ-Method

The derivative of a function expresses its rate of change with respect to an independent variable. The derivative is also the slope of the tangent line to the curve.

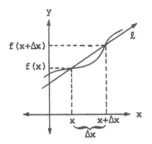

Fig. 5.1

Consider the graph of the function f in Fig. 5.1. Choosing a point x and a point x + Δx (where Δx denotes a small distance on the x-axis) we can obtain both, f(x) and f(x+Δx). Drawing a tangent line, ℓ, of the curve through the points f(x) and f(x+Δx), we can measure the rate of change of this line. As we let the distance, Δx, approach zero, then

$$\lim_{\Delta x \to 0} \frac{f(x+\Delta x)-f(x)}{\Delta x}$$

becomes the instantaneous rate of change of the function or the derivative.

We denote the derivative of the function f to be f'. So we have

$$f'(x) = \lim_{\Delta x \to 0} \frac{f(x+\Delta x)-f(x)}{\Delta x}$$

If y = f(x), some common notations for the derivative are

$$y' = f'(x)$$

$$\frac{dy}{dx} = f'(x)$$

$$D_x y = f'(x) \text{ or } Df = f'$$

17

Additional Problem Solving Examples:

 Find the slope of the following curve at the given point, using the Δ-method

$$y = 3x^2 - 2x + 4 \text{ at } (1,5)$$

A The slope of a given curve at a specified point is the derivative, in this case $\dfrac{\Delta y}{\Delta x}$, evaluated at that point.

From the Δ-method we know that:

$$\frac{\Delta y}{\Delta x} = \frac{f(x + \Delta x) - f(x)}{\Delta x} \quad .$$

For the curve $y = 3x^2 - 2x + 4$, we find:

$$\frac{\Delta y}{\Delta x} = \frac{3(x + \Delta x)^2 - 2(x + \Delta x) + 4 - (3x^2 - 2x + 4)}{\Delta x}$$

$$= \frac{3x^2 + 6x\Delta x + 3(\Delta x)^2 - 2x - 2\Delta x + 4 - 3x^2 + 2x - 4}{\Delta x}$$

$$= \frac{6x\Delta x + 3(\Delta x)^2 - 2\Delta x}{\Delta x}$$

$$= 6x + 3\Delta x - 2.$$

$$\lim_{\Delta x \to 0} \frac{\Delta y}{\Delta x} = \lim_{\Delta x \to 0} 6x + 3\Delta x - 2 = 6x - 2.$$

At $(1,5)\ \dfrac{\Delta y}{\Delta x} = 4$ is the required slope.

Find the average rate of change, by the Δ process, for:

$$y = \frac{1}{x} .$$

A $y = f(x) = \dfrac{1}{x}$

The average rate of change is defined to be

$$\frac{\Delta y}{\Delta x} \text{ with } \Delta y = f(x + \Delta x) - f(x).$$

Since

$$f(x) = \frac{1}{x}, f(x + \Delta x) = \frac{1}{x + \Delta x} ,$$

and

$$\Delta y = \frac{1}{x + \Delta x} - \frac{1}{x} = \frac{x - (x + \Delta x)}{x\,(x + \Delta x)}$$

$$= \frac{-\Delta x}{x\,(x + \Delta x)} .$$

Now,

$$\frac{\Delta y}{\Delta x} = \frac{-\Delta x}{x\,(x + \Delta x)\,\Delta x} = -\frac{1}{x\,(x + \Delta x)} .$$

Therefore, the average rate of change is $\dfrac{-1}{x\,(x + \Delta x)} .$

CHAPTER 4

Simplest Cases

Now let us see how, on first principles, we can differentiate some simple algebraical expressions.

Case 1.

Let us begin with the simple expression $y = x^2$. Now remember that the fundamental notion about the calculus is the idea of *growing*. Mathematicians call it *varying*. Now as y and x^2 are equal to one another, it is clear that if x grows, x^2 will also grow. And if x^2 grows, then y will also grow. What we have got to find out is the proportion between the growing of y and the growing of x. In other words, our task is to find out the ratio between dy and dx, or, in brief, to find the value of $\dfrac{dy}{dx}$.

Let x, then, grow a little bit bigger and become $x + dx$; similarly, y will grow a bit bigger and will become $y + dy$. Then, clearly, it will still be true that the enlarged y will be equal to the square of the enlarged x. Writing this down, we have:

$$y + dy = (x + dx)^2$$

Doing the squaring we get:

$$y + dy = x^2 + 2x \cdot dx + (dx)^2$$

What does $(dx)^2$ mean? Remember that dx meant a bit—a little bit—of x. Then $(dx)^2$ will mean a little bit of a little bit of

x^2; that is, as explained above, it is a small quantity of the second order of smallness. It may therefore be discarded as quite negligible in comparison with the other terms. Leaving it out, we then have:

$$y + dy = x^2 + 2x \cdot dx$$

Now $y = x^2$; so let us subtract this from the equation and we have left

$$dy = 2x \cdot dx$$

Dividing across by dx, we find

$$\frac{dy}{dx} = 2x$$

Now *this** is what we set out to find. The ratio of the growing of y to the growing of x is, in the case before us, found to be $2x$.

Numerical Example.

Suppose $x = 100$ and therefore $y = 10,000$. Then let x grow till it becomes 101 (that is, let $dx = 1$). Then the enlarged y will be $101 \times 101 = 10,201$. But if we agree that we may ignore small quantities of the second order, 1 may be rejected as compared with 10,000; so we may round off the enlarged y to 10,200; y has

---

*N.B.—This ratio $\frac{dy}{dx}$ is the result of differentiating y with respect to x. Differentiating means finding the derivative. Suppose we had some other function of x, as, for example, $u = 7x^2 + 3$. Then if we were told to differentiate this with respect to x, we should have to find $\frac{du}{dx}$, or, what is the same thing, $\frac{d(7x^2 + 3)}{dx}$. On the other hand, we may have a case in which time was the independent variable, such as this: $y = b + \frac{1}{2}at^2$. Then, if we were told to differentiate it, that means we must find its derivative with respect to t. So that then our business would be to try to find $\frac{dy}{dt}$, that is, to find $\frac{d\left(b + \frac{1}{2}at^2\right)}{dt}$.

grown from 10,000 to 10,200; the bit added on is dy, which is therefore 200.

$\dfrac{dy}{dx} = \dfrac{200}{1} = 200$. According to the algebra-working of the previous paragraph, we find $\dfrac{dy}{dx} = 2x$. And so it is; for $x = 100$ and $2x = 200$.

But, you will say, we neglected a whole unit.

Well, try again, making dx a still smaller bit.

Try $dx = \frac{1}{10}$. Then $x + dx = 100.1$, and

$$(x + dx)^2 = 100.1 \times 100.1 = 10,020.01$$

Now the last figure 1 is only one-millionth part of the 10,000, and is utterly negligible; so we may take 10,020 without the little decimal at the end. And this makes $dy = 20$; and $\dfrac{dy}{dx} = \dfrac{20}{0.1} = 200$, which is still the same as $2x$.

Case 2.

Try differentiating $y = x^3$ in the same way.

We let y grow to $y + dy$, while x grows to $x + dx$.

Then we have

$$y + dy = (x + dx)^3$$

Doing the cubing we obtain

$$y + dy = x^3 + 3x^2 \cdot dx + 3x(dx)^2 + (dx)^3$$

Now we know that we may neglect small quantities of the second and third orders; since, when dy and dx are both made infinitesimal small, $(dx)^2$ and $(dx)^3$ will become infinitesimal smaller by comparison. So, regarding them as negligible, we have left :

$$y + dy = x^3 + 3x^2 \cdot dx$$

But $y = x^3$; and, subtracting this, we have:

$$dy = 3x^2 \cdot dx$$

and $\qquad\qquad \dfrac{dy}{dx} = 3x^2$

22

Case 3.

Try differentiating $y = x^4$. Starting as before by letting both y and x grow a bit, we have:

$$y + dy = (x + dx)^4$$

Working out the raising to the fourth power, we get

$$y + dy = x^4 + 4x^3dx + 6x^2(dx)^2 + 4x(dx)^3 + (dx)^4$$

Then, striking out the terms containing all the higher powers of dx, as being negligible by comparison, we have

$$y + dy = x^4 + 4x^3dx$$

Subtracting the original $y = x^4$, we have left

$$dy = 4x^3dx, \quad \text{and} \quad \frac{dy}{dx} = 4x^3$$

Now all these cases are quite easy. Let us collect the results to see if we can infer any general rule. Put them in two columns, the values of y in one and the corresponding values found for $\frac{dy}{dx}$ in the other: thus

y	$\dfrac{dy}{dx}$
x^2	$2x$
x^3	$3x^2$
x^4	$4x^3$

Just look at these results: the operation of differentiating appears to have had the effect of diminishing the power of x by 1 (for example in the last case reducing x^4 to x^3), and at the same time multiplying by a number (the same number in fact which originally appeared as the power). Now, when you have once seen this, you might easily conjecture how the others will run. You would expect that differentiating x^5 would give $5x^4$, or differentiating x^6 would give $6x^5$. If you hesitate, try one of these, and see whether the conjecture comes right.

Try $y = x^5$.

Then $y + dy = (x + dx)^5 = x^5 + 5x^4dx + 10x^3(dx)^2 + 10x^2(dx)^3$
$$+ 5x(dx)^4 + (dx)^5.$$

Neglecting all the terms containing small quantities of the higher orders, we have left

$$y + dy = x^5 + 5x^4dx$$

and subtracting $y = x^5$ leaves us

$$dy = 5x^4dx$$

whence $\quad \dfrac{dy}{dx} = 5x^4$, exactly as we supposed.

Following out logically our observation, we should conclude that if we want to deal with any higher power—call it x^n—we could tackle it in the same way.

Let $\qquad\qquad y = x^n$

then we should expect to find that

$$\frac{dy}{dx} = nx^{n-1}$$

For example, let $n = 8$, then $y = x^8$; and differentiating it would give $\dfrac{dy}{dx} = 8x^7$.

And, indeed, the rule that differentiating x^n gives as the result nx^{n-1} is true for all cases where n is a whole number and positive. [Expanding $(x + dx)^n$ by the binomial theorem will at once show this.] But the question whether it is true for cases where n has negative or fractional values requires further consideration.

Case of a Negative Exponent.
Let $y = x^{-2}$. Then proceed as before:

$$y + dy = (x + dx)^{-2}$$

$$= x^{-2}\left(1 + \frac{dx}{x}\right)^{-2}$$

Expanding this by the binomial theorem, we get

$$= x^{-2}\left[1 - \frac{2dx}{x} + \frac{2(2+1)}{1 \times 2}\left(\frac{dx}{x}\right)^2 - \cdots\right]$$

$$= x^{-2} - 2x^{-3} \cdot dx + 3x^{-4}(dx)^2 - 4x^{-5}(dx)^3 + \text{etc.}$$

So, neglecting the small quantities of higher orders of smallness, we have:

$$y + dy = x^{-2} - 2x^{-3} \cdot dx$$

Subtracting the original $y = x^{-2}$, we find

$$dy = -2x^{-3}dx$$

$$\frac{dy}{dx} = -2x^{-3}$$

And this is still in accordance with the rule inferred above.

Case of a Fractional Exponent.

Let $y = x^{\frac{1}{2}}$. Then, as before,

$$y + dy = (x + dx)^{\frac{1}{2}} = x^{\frac{1}{2}}\left(1 + \frac{dx}{x}\right)^{\frac{1}{2}} = \sqrt{x}\left(1 + \frac{dx}{x}\right)^{\frac{1}{2}}$$

$$= \sqrt{x} + \frac{1}{2}\frac{dx}{\sqrt{x}} - \frac{1}{8}\frac{(dx)^2}{x\sqrt{x}} + \begin{array}{l}\text{terms with higher} \\ \text{powers of } dx.\end{array}$$

Subtracting the original $y = x^{\frac{1}{2}}$, and neglecting higher powers we have left:

$$dy = \frac{1}{2} \frac{dx}{\sqrt{x}} = \frac{1}{2} x^{-\frac{1}{2}} \cdot dx$$

and $\dfrac{dy}{dx} = \dfrac{1}{2} x^{-\frac{1}{2}}$. This agrees with the general rule.

Summary. Let us see how far we have got. We have arrived at the following rule: To differentiate x^n, multiply it by the exponent and reduce the exponent by one, so giving us nx^{n-1} as the result.

Exercises I

(See Answers on page 296)

Differentiate the following:

(1) $y = x^{13}$

(2) $y = x^{-\frac{3}{2}}$

(3) $y = x^{2a}$

(4) $u = t^{2.4}$

(5) $z = \sqrt[3]{u}$

(6) $y = \sqrt[3]{x^{-5}}$

(7) $u = \sqrt[5]{\dfrac{1}{x^8}}$

(8) $y = 2x^a$

(9) $y = \sqrt[q]{x^3}$

(10) $y = \sqrt[n]{\dfrac{1}{x^m}}$

You have now learned how to differentiate powers of x. How easy it is!

CHAPTER 5

Next Stage—What To Do With Constants

In our equations we have regarded x as growing, and as a result of x being made to grow y also changed its value and grew. We usually think of x as a quantity that we can vary; and, regarding the variation of x as a sort of *cause*, we consider the resulting variation of y as an *effect*. In other words, we regard the value of y as depending on that of x. Both x and y are variables, but x is the one that we operate upon, and y is the "dependent variable". In all the preceding chapters we have been trying to find out rules for the proportion which the dependent variation in y bears to the variation independently made in x.

Our next step is to find out what effect on the process of differentiating is caused by the presence of *constants*, that is, of numbers which don't change when x or y changes its value.

Added Constants.

Let us begin with a simple case of an added constant, thus:

Let $$y = x^3 + 5$$

Just as before, let us suppose x to grow to $x + dx$ and y to grow to $y + dy$.

Then: $$y + dy = (x + dx)^3 + 5$$
$$= x^3 + 3x^2 dx + 3x(dx)^2 + (dx)^3 + 5$$

Neglecting the small quantities of higher orders, this becomes

$$y + dy = x^3 + 3x^2 \cdot dx + 5$$

Subtract the original $y = x^3 + 5$, and we have left:

$$dy = 3x^2dx$$

$$\frac{dy}{dx} = 3x^2$$

So the 5 has quite disappeared. It added nothing to the growth of x, and does not enter into the derivative. If we had put 7, or 700, or any other number, instead of 5, it would have disappeared. So if we take the letter a, or b, or c to represent any constant, it will simply disappear when we differentiate.

If the additional constant had been of negative value, such as -5 or $-b$, it would equally have disappeared.

Multiplied Constants.

Take as a simple experiment this case:

Let $y = 7x^2$

Then on proceeding as before we get:

$$y + dy = 7(x + dx)^2$$
$$= 7\{x^2 + 2x \cdot dx + (dx)^2\}$$
$$= 7x^2 + 14x \cdot dx + 7(dx)^2$$

Then, subtracting the original $y = 7x^2$, and neglecting the last term, we have

$$dy = 14x \cdot dx$$

$$\frac{dy}{dx} = 14x$$

Let us illustrate this example by working out the graphs of the equations $y = 7x^2$ and $\frac{dy}{dx} = 14x$, by assigning to x a set of successive values, 0, 1, 2, 3, etc., and finding the corresponding values of y and of $\frac{dy}{dx}$.

These values we tabulate as follows:

x	0	1	2	3	4	5	−1	−2	−3
y	0	7	28	63	112	175	7	28	63
$\dfrac{dy}{dx}$	0	14	28	42	56	70	−14	−28	−42

Now plot these values to some convenient scale, and we obtain the two curves, Figs. 6 and 6*a*.

Carefully compare the two figures, and verify by inspection that the height of the ordinate of the derived curve, Fig. 6*a*, is proportional to the *slope* of the original curve, Fig. 6, at the corresponding value of *x*. To the left of the origin, where the original curve slopes negatively (that is, downward from left to right) the corresponding ordinates of the derived curve are negative.

Now, if we look back at previous pages, we shall see that simply differentiating x^2 gives us $2x$. So that the derivative of $7x^2$ is just 7 times as big as that of x^2. If we had taken $8x^2$, the derivative would have come out eight times as great as that of x^2. If we put $y = ax^2$, we shall get

$$\frac{dy}{dx} = a \times 2x$$

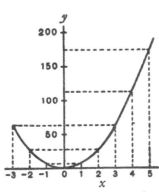

FIG. 6. Graph of $y = 7x^2$.

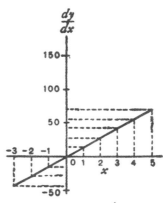

FIG. 6a. Graph of $\dfrac{dy}{dx} = 14x$.

29

If we had begun with $y = ax^n$, we should have had

$$\frac{dy}{dx} = a \times nx^{n-1}$$

So that any mere multiplication by a constant reappears as a mere multiplication when the thing is differentiated. And what is true about multiplication is equally true about *division:* for if, in the example above, we had taken as the constant $\frac{1}{7}$ instead of 7, we should have had the same $\frac{1}{7}$ come out in the result after differentiation.

Some Further Examples.

The following further examples, fully worked out, will enable you to master completely the process of differentiation as applied to ordinary algebraical expressions, and enable you to work out by yourself the examples given at the end of this chapter.

(1) Differentiate $y = \dfrac{x^5}{7} - \dfrac{3}{5}$

$-\dfrac{3}{5}$ is an added constant and vanishes.

We may then write at once

$$\frac{dy}{dx} = \frac{1}{7} \times 5 \times x^{5-1}$$

or
$$\frac{dy}{dx} = \frac{5}{7}\, x^4$$

(2) Differentiate $y = a\sqrt{x} - \dfrac{1}{2}\sqrt{a}$

The term $-\dfrac{1}{2}\sqrt{a}$ vanishes, being an added constant; and as $a\sqrt{x}$, in the index form, is written $ax^{\frac{1}{2}}$, we have

$$\frac{dy}{dx} = a \times \frac{1}{2} \times x^{\frac{1}{2}-1} = \frac{a}{2} \times x^{-\frac{1}{2}}$$

or
$$\frac{dy}{dx} = \frac{a}{2\sqrt{x}}$$

(3) the volume of a cylinder of radius r and height h is given by the formula $V = \pi r^2 h$.

Find the rate of variation of volume with the radius when $r = 5.5$ in. and $h = 20$ in. If $r = h$, find the dimensions of the cylinder so that a change of 1 in. in radius causes a change of 400 cubic inches in the volume. The rate of variation of V with regard to r is

$$\frac{dV}{dr} = 2\pi r h$$

If $r = 5.5$ in. and $h = 20$ in. this becomes 691.2. It means that a change of radius of 1 inch will cause a change of volume of 691.2 cubic inches. This can be easily verified, for the volumes with $r = 5$ and $r = 6$ are 1570.8 cubic inches and 2262 cubic inches respectively, and $2262 - 1570.8 = 691.2$.

Also, if $h = r$, and h remains constant,

$$\frac{dV}{dr} = 2\pi r^2 = 400 \quad \text{and} \quad r = h = \sqrt{\frac{400}{2\pi}} = 7.98 \text{ in.}$$

If, however, $h = r$ and varies with r, then

$$\frac{dV}{dr} = 3\pi r^2 = 400 \quad \text{and} \quad r = h = \sqrt{\frac{400}{3\pi}} = 6.51 \text{ in.}$$

(4) The reading θ of a Féry's Radiation pyrometer is related to the centigrade temperature t of the observed body by the relation $\theta/\theta_1 = (t/t_1)^4$, where θ_1 is the reading corresponding to a known temperature t_1 of the observed body.

Compare the sensitivity of the pyrometer at temperatures 800° C., 1000° C., 1200° C., given that it read 25 when the temperature was 1000° C.

The sensitivity is the rate of variation of the reading with the temperature, that is, $\dfrac{d\theta}{dt}$. The formula may be written

$$\theta = \frac{\theta_1}{t_1^{\,4}}\, t^4 = \frac{25 t^4}{1000^4}$$

and we have $\quad \dfrac{d\theta}{dt} = \dfrac{100 t^3}{1000^4} = \dfrac{t^3}{10,000,000,000}$

31

When $t = 800$, 1000 and 1200, we get $\dfrac{d\theta}{dt} = 0.0512$, 0.1 and 0.1728 respectively.

The sensitivity is approximately doubled from $800°$ to $1000°$, and becomes three-quarters as great again up to $1200°$.

Exercises II

(See answers on page 296)

Differentiate the following:

(1) $y = ax^3 + 6$

(2) $y = 13x^{\frac{3}{2}} - c$

(3) $y = 12x^{\frac{1}{2}} + c^{\frac{1}{2}}$

(4) $y = c^{\frac{1}{2}}x^{\frac{1}{2}}$

(5) $u = \dfrac{az^n - 1}{c}$

(6) $y = 1.18t^2 + 22.4$

Make up some other examples for yourself, and try your hand at differentiating them.

(7) If l_t and l_0 be the lengths of a rod of iron at the temperatures $t°$ C. and $0°$ C. respectively, then $l_t = l_0 (1 + 0.000012t)$. Find the change of length of the rod per degree centigrade.

(8) It has been found that if c be the candle power of an incandescent electric lamp, and V be the voltage, $c = aV^b$, where a and b are constants.

Find the rate of change of the candle power with the voltage, and calculate the change of candle power per volt at 80, 100 and 120 volts in the case of a lamp for which $a = 0.5 \times 10^{-10}$ and $b = 6$.

(9) The frequency n of vibration of a string of diameter D, length L and specific gravity σ, stretched with a force T, is given by

$$n = \frac{1}{DL} \sqrt{\frac{gT}{\pi\sigma}}$$

Find the rate of change of the frequency when D, L, σ and T are varied singly.

(10) The greatest external pressure P which a tube can support without collapsing is given by

$$P = \left(\frac{2E}{1 - \sigma^2} \right) \frac{t^3}{D^3}$$

where E and σ are constants, t is the thickness of the tube and D is its diameter. (This formula assumes that $4t$ is small compared to D.)

Compare the rate at which P varies for a small change of thickness and for a small change of diameter taking place separately.

(11) Find, from first principles, the rate at which the following vary with respect to a change in radius:

(a) the circumference of a circle of radius r;
(b) the area of a circle of radius r;
(c) the lateral area of a cone of slant dimension l;
(d) the volume of a cone of radius r and height h;
(e) the area of a sphere of radius r;
(f) the volume of a sphere of radius r.

CHAPTER 6

Differentiating Sums, Differences, Products, & Quotients of Functions

We have learned how to differentiate simple algebraical functions such as $x^2 + c$ or ax^4, and we have now to consider how to tackle the *sum* of two or more functions.

For instance, let

$$y = (x^2 + c) + (ax^4 + b)$$

what will its $\dfrac{dy}{dx}$ be? How are we to go to work on this new job?

The answer to this question is quite simple: just differentiate them, one after the other, thus:

$$\frac{dy}{dx} = 2x + 4ax^3$$

If you have any doubt whether this is right, try a more general case, working it by first principles. And this is the way.

Let $y = u + v$, where u is any function of x, and v any other function of x. Then, letting x increase to $x + dx$, y will increase to $y + dy$; and u will increase to $u + du$; and v to $v + dv$.

And we shall have:

$$y + dy = u + du + v + dv$$

34

Subtracting the original $y = u + v$, we get

$$dy = du + dv$$

and dividing through by dx, we get:

$$\frac{dy}{dx} = \frac{du}{dx} + \frac{dv}{dx}$$

This justifies the procedure. You differentiate each function separately and add the results. So if now we take the example of the preceding paragraph, and put in the values of the two functions, we shall have, using the notation shown,

$$\frac{dy}{dx} = \frac{d(x^2 + c)}{dx} + \frac{d(ax^4 + b)}{dx}$$

$$= 2x \qquad + 4ax^3$$

exactly as before.

If there were three functions of x, which we may call u, v and w, so that

$$y = u + v + w$$

then

$$\frac{dy}{dx} = \frac{du}{dx} + \frac{dv}{dx} + \frac{dw}{dx}$$

As for the rule about *subtraction*, it follows at once; for if the function v had itself had a negative sign, its derivative would also be negative; so that by differentiating

$$y = u - v$$

we should get

$$\frac{dy}{dx} = \frac{du}{dx} - \frac{dv}{dx}$$

But when we come to do with *Products*, the thing is not quite so simple.

Suppose we were asked to differentiate the expression

$$y = (x^2 + c) \times (ax^4 + b)$$

35

what are we to do? The result will certainly *not* be $2x \times 4ax^3$; for it is easy to see that neither $c \times ax^4$, nor $x^2 \times b$, would have been taken into that product.

Now there are two ways in which we may go to work.

First Way. Do the multiplying first, and, having worked it out, then differentiate.

Accordingly, we multiply together $x^2 + c$ and $ax^4 + b$.

This gives $ax^6 + acx^4 + bx^2 + bc$.

Now differentiate, and we get:

$$\frac{dy}{dx} = 6ax^5 + 4acx^3 + 2bx$$

Second Way. Go back to first principles, and consider the equation

$$y = u \times v$$

where u is one function of x, and v is any other function of x. Then, if x grows to be $x + dx$; and y to $y + dy$; and u becomes $u + du$; and v becomes $v + dv$, we shall have:

$$y + dy = (u + du) \times (v + dv)$$

$$= u \cdot v + u \cdot dv + v \cdot du + du \cdot dv$$

Now $du \cdot dv$ is a small quantity of the second order of smallness, and therefore in the limit may be discarded, leaving

$$y + dy = u \cdot v + u \cdot dv + v \cdot du$$

Then, subtracting the original $y = u \cdot v$, we have left

$$dy = u \cdot dv + v \cdot du$$

and, dividing through by dx, we get the result:

$$\frac{dy}{dx} = u\frac{dv}{dx} + v\frac{du}{dx}$$

This shows that our instructions will be as follows: *To differentiate the product of two functions, multiply each function by the derivative of the other, and add together the two products so obtained.*

You should note that this process amounts to the following:

36

Treat u as constant while you differentiate *v*; then treat *v* as constant while you differentiate *u*; and the whole derivative $\dfrac{dy}{dx}$ will be the sum of the results of these two treatments.

Now, having found this rule, apply it to the concrete example which was considered above.

We want to differentiate the product

$$(x^2 + c) \times (ax^4 + b)$$

Call $(x^2 + c) = u;$ and $(ax^4 + b) = v$

Then, by the general rule just established, we may write:

$$\frac{dy}{dx} = (x^2 + c)\frac{d(ax^4 + b)}{dx} + (ax^4 + b)\frac{d(x^2 + c)}{dx}$$

$$= (x^2 + c)4ax^3 \qquad + (ax^4 + b)2x$$

$$= 4ax^5 + 4acx^3 \qquad + 2ax^5 + 2bx$$

$$\frac{dy}{dx} = 6ax^5 + 4acx^3 \qquad + 2bx$$

Exactly as before.

Lastly, we have to differentiate *quotients*.

Think of this example, $y = \dfrac{bx^5 + c}{x^2 + a}$. In such a case it is no use to try to work out the division beforehand, because $x^2 + a$ will not divide into $bx^5 + c$, neither have they any common factor. So there is nothing for it but to go back to first principles, and find a rule.

So we will put $y = \dfrac{u}{v}$

where *u* and *v* are two different functions of the independent variable *x*. Then, when *x* becomes $x + dx$, *y* will become $y + dy$; and *u* will become $u + du$; and *v* will become $v + dv$. So then

$$y + dy = \frac{u + du}{v + dv}$$

37

Now perform the algebraic division, thus:

$$v + dv \;\Big|\; u + du \;\Big|\; \frac{u}{v} + \frac{du}{v} - \frac{u \cdot dv}{v^2}$$

$$u + \frac{u \cdot dv}{v}$$

$$\overline{\qquad\qquad}$$

$$du - \frac{u \cdot dv}{v}$$

$$du + \frac{du \cdot dv}{v}$$

$$\overline{\qquad\qquad}$$

$$-\frac{u \cdot dv}{v} - \frac{du \cdot dv}{v}$$

$$-\frac{u \cdot dv}{v} - \frac{u \cdot dv \cdot dv}{v^2}$$

$$\overline{\qquad\qquad}$$

$$-\frac{du \cdot dv}{v} + \frac{u \cdot dv \cdot dv}{v^2}$$

As both these remainders are small quantities of the second order, they may be neglected, and the division may stop here, since any further remainders would be of still smaller magnitudes.

So we have got:

$$y + dy = \frac{u}{v} + \frac{du}{v} - \frac{u \cdot dv}{v^2}$$

which may be written

$$= \frac{u}{v} + \frac{v \cdot du - u \cdot dv}{v^2}$$

Now subtract the original $y = \dfrac{u}{v}$, and we have left:

$$dy = \frac{v \cdot du - u \cdot dv}{v^2}$$

whence

$$\frac{dy}{dx} = \frac{v \dfrac{du}{dx} - u \dfrac{dv}{dx}}{v^2}$$

This gives us our instructions as to *how to differentiate a quotient of* two functions. *Multiply the divisor function by the derivative of the dividend function; then multiply the dividend function by the derivative of the divisor function; and subtract the latter product from the former. Lastly, divide the difference by the square of the divisor function.*

Going back to our example $y = \dfrac{bx^5 + c}{x^2 + a}$

write $\qquad bx^5 + c = u; \quad \text{and} \quad x^2 + a = v$

Then $\quad \dfrac{dy}{dx} = \dfrac{(x^2 + a)\dfrac{d(bx^5 + c)}{dx} - (bx^5 + c)\dfrac{d(x^2 + a)}{dx}}{(x^2 + a)^2}$

$$= \dfrac{(x^2 + a)(5bx^4) - (bx^5 + c)(2x)}{(x^2 + a)^2}$$

$$\dfrac{dy}{dx} = \dfrac{3bx^6 + 5abx^4 - 2cx}{(x^2 + a)^2}$$

The working out of quotients is often tedious, but there is nothing difficult about it.

Some further examples fully worked out are given hereafter.

(1) Differentiate $\quad y = \dfrac{a}{b^2}x^3 - \dfrac{a^2}{b}x + \dfrac{a^2}{b^2}$

Being a constant, $\dfrac{a^2}{b^2}$ vanishes, and we have

$$\frac{dy}{dx} = \frac{a}{b^2} \times 3 \times x^{3-1} - \frac{a^2}{b} \times 1 \times x^{1-1}$$

But $x^{1-1} = x^0 = 1$; so we get:

$$\frac{dy}{dx} = \frac{3a}{b^2}x^2 - \frac{a^2}{b}$$

(2) Differentiate $\quad y = 2a\sqrt{bx^3} - \dfrac{3b\sqrt[3]{a}}{x} - 2\sqrt{ab}$

Putting x in the exponent form, we get

$$y = 2a\sqrt{b}\, x^{\frac{3}{2}} - 3b\sqrt[3]{a}\, x^{-1} - 2\sqrt{ab}$$

Now $\quad \dfrac{dy}{dx} = 2a\sqrt{b} \times \tfrac{3}{2} \times x^{\frac{3}{2}-1} - 3b\sqrt[3]{a} \times (-1) \times x^{-1-1}$

or, $\quad \dfrac{dy}{dx} = 3a\sqrt{bx} + \dfrac{3b\sqrt[3]{a}}{x^2}$

(3) Differentiate $\quad z = 1.8\sqrt[3]{\dfrac{1}{\theta^2}} - \dfrac{4.4}{\sqrt[5]{\theta}} - 27.$

This may be written: $z = 1.8\theta^{-\frac{2}{3}} - 4.4\theta^{-\frac{1}{5}} - 27.$
The 27 vanishes, and we have

$$\dfrac{dz}{d\theta} = 1.8 \times \left(-\tfrac{2}{3}\right)\theta^{-\frac{2}{3}-1} - 4.4 \times \left(-\tfrac{1}{5}\right)\theta^{-\frac{1}{5}-1}$$

or, $\quad \dfrac{dz}{d\theta} = -1.2\theta^{-\frac{5}{3}} + 0.88\theta^{-\frac{6}{5}}$

or, $\quad \dfrac{dz}{d\theta} = \dfrac{0.88}{\sqrt[5]{\theta^6}} - \dfrac{1.2}{\sqrt[3]{\theta^5}}$

(4) Differentiate $\quad v = (3t^2 - 1.2t + 1)^3.$
A direct way of doing this will be explained later; but we can nevertheless manage it now without any difficulty.
Developing the cube, we get

$$v = 27t^6 - 32.4t^5 + 39.96t^4 - 23.328t^3 + 13.32t^2 - 3.6t + 1$$

hence

$$\dfrac{dv}{dt} = 162t^5 - 162t^4 + 159.84t^3 - 69.984t^2 + 26.64t - 3.6.$$

(5) Differentiate $\quad y = (2x - 3)(x + 1)^2.$

$$\frac{dy}{dx} = (2x - 3)\frac{d[(x + 1)(x + 1)]}{dx} + (x + 1)^2\frac{d(2x - 3)}{dx}$$

$$= (2x - 3)\left[(x + 1)\frac{d(x + 1)}{dx} + (x + 1)\frac{d(x + 1)}{dx}\right]$$

$$+ (x + 1)^2\frac{d(2x - 3)}{dx}$$

$$= 2(x + 1)[(2x - 3) + (x + 1)] = 2(x + 1)(3x - 2)$$

or, more simply, multiply out and then differentiate.

(6) Differentiate $y = 0.5x^3(x - 3)$.

$$\frac{dy}{dx} = 0.5\left[x^3\frac{d(x - 3)}{dx} + (x - 3)\frac{d(x^3)}{dx}\right]$$

$$= 0.5[x^3 + (x - 3) \times 3x^2] = 2x^3 - 4.5x^2$$

Same remarks as for preceding example.

(7) Differentiate $w = \left(\theta + \dfrac{1}{\theta}\right)\left(\sqrt{\theta} + \dfrac{1}{\sqrt{\theta}}\right)$

This may be written

$$w = (\theta + \theta^{-1})(\theta^{\frac{1}{2}} - \theta^{-\frac{1}{2}})$$

$$\frac{dw}{d\theta} = (\theta + \theta^{-1})\frac{d(\theta^{\frac{1}{2}} + \theta^{-\frac{1}{2}})}{d\theta} + (\theta^{\frac{1}{2}} + \theta^{-\frac{1}{2}})\frac{d(\theta + \theta^{-1})}{d\theta}$$

$$= (\theta + \theta^{-1})\left(\frac{1}{2}\theta^{-\frac{1}{2}} - \frac{1}{2}\theta^{-\frac{3}{2}}\right) + (\theta^{\frac{1}{2}} + \theta^{-\frac{1}{2}})(1 - \theta^{-2})$$

$$= \frac{1}{2}(\theta^{\frac{1}{2}} + \theta^{-\frac{3}{2}} - \theta^{-\frac{1}{2}} - \theta^{-\frac{5}{2}}) + (\theta^{\frac{1}{2}} + \theta^{-\frac{1}{2}} - \theta^{-\frac{3}{2}} - \theta^{-\frac{5}{2}})$$

$$= \frac{3}{2}\left(\sqrt{\theta} - \frac{1}{\sqrt{\theta^5}}\right) + \frac{1}{2}\left(\frac{1}{\sqrt{\theta}} - \frac{1}{\sqrt{\theta^3}}\right)$$

This, again, could be obtained more simply by multiplying the two factors first, and differentiating afterwards. This is not, however, always possible; see, for instance, Example 8 in Chapter 15, in which the rule for differentiating a product *must* be used.

(8) Differentiate $y = \dfrac{a}{1 + a\sqrt{x} + a^2 x}$

$$\frac{dy}{dx} = \frac{\left(1 + ax^{\frac{1}{2}} + a^2 x\right) \times 0 - a\dfrac{d\left(1 + ax^{\frac{1}{2}} + a^2 x\right)}{dx}}{\left(1 + a\sqrt{x} + a^2 x\right)^2}$$

$$= -\frac{a\left(\frac{1}{2}ax^{-\frac{1}{2}} + a^2\right)}{\left(1 + ax^{\frac{1}{2}} + a^2 x\right)^2}$$

(9) Differentiate $y = \dfrac{x^2}{x^2 + 1}$.

$$\frac{dy}{dx} = \frac{(x^2 + 1)2x - x^2 \times 2x}{(x^2 + 1)^2} = \frac{2x}{(x^2 + 1)^2}$$

(10) Differentiate $y = \dfrac{a + \sqrt{x}}{a - \sqrt{x}}$

In the exponent form, $y = \dfrac{a + x^{\frac{1}{2}}}{a - x^{\frac{1}{2}}}$

$$\frac{dy}{dx} = \frac{\left(a - x^{\frac{1}{2}}\right)\left(\frac{1}{2}x^{-\frac{1}{2}}\right) - \left(a + x^{\frac{1}{2}}\right)\left(-\frac{1}{2}x^{-\frac{1}{2}}\right)}{\left(a - x^{\frac{1}{2}}\right)^2} = \frac{a - x^{\frac{1}{2}} + a + x^{\frac{1}{2}}}{2\left(a - x^{\frac{1}{2}}\right)^2 x^{\frac{1}{2}}}$$

hence $\qquad \dfrac{dy}{dx} = \dfrac{a}{\left(a - \sqrt{x}\right)^2 \sqrt{x}}$

(11) Differentiate $\theta = \dfrac{1 - a\sqrt[3]{t^2}}{1 + a\sqrt[2]{t^3}}$

Now $\qquad \theta = \dfrac{1 - at^{\frac{2}{3}}}{1 + at^{\frac{3}{2}}}$

$$\frac{d\theta}{dt} = \frac{\left(1 + at^{\frac{3}{2}}\right)\left(-\frac{2}{3}at^{-\frac{1}{3}}\right) - \left(1 - at^{\frac{2}{3}}\right) \times \frac{3}{2}at^{\frac{1}{2}}}{\left(1 + at^{\frac{3}{2}}\right)^2}$$

$$= \frac{5a^2\sqrt[6]{t^7} - \frac{4a}{\sqrt[3]{t}} - 9a\sqrt[2]{t}}{6\left(1 + a\sqrt[2]{t^3}\right)^2}$$

(12) A reservoir of square cross-section has sides sloping at an angle of 45° with the vertical. The side of the bottom is p feet in length, and water flows in the reservoir at the rate of c cubic feet per minute. Find an expression for the rate at which the surface of the water is rising at the instant its depth is h feet. Calculate this rate when $p = 17$, $h = 4$ and $c = 35$.

The volume of a frustum of pyramid of height H, and of bases A and a, is $V = \frac{H}{3}\left(A + a + \sqrt{Aa}\right)$. It is easily seen that, the slope being 45°, for a depth of h, the length of the side of the upper square surface of water is $(p + 2h)$ feet; thus $A = p^2$, $a = (p + 2h)^2$ and the volume of the water is

$$\tfrac{1}{3}h\{p^2 + p(p + 2h) + (p + 2h)^2\} \text{ cubic feet}$$

$$= p^2h + 2ph^2 + \tfrac{4}{3}h^3 \text{ cubic feet}$$

Now, if t be the time in minutes taken for this volume of water to flow in,

$$ct = p^2h + 2ph^2 + \tfrac{4}{3}h^3$$

From this relation we have the rate at which h increases with t, that is $\frac{dh}{dt}$, but as the above expression is in the form, $t =$ function of h, rather than $h =$ function of t, it will be easier to find $\frac{dt}{dh}$ and then invert the result, for

$$\frac{dt}{dh} \times \frac{dh}{dt} = 1$$

43

Hence, since c and p are constants, and

$$ct = p^2h + 2ph^2 + \tfrac{4}{3}h^3$$

$$c\frac{dt}{dh} = p^2 + 4ph + 4h^2 = (p + 2h)^2$$

so that $\dfrac{dh}{dt} = \dfrac{c}{(p + 2h)^2}$, which is the required expression.

When $p = 17$, $h = 4$ and $c = 35$; this becomes 0.056 feet per minute.

(13) The absolute pressure, in atmospheres, P, of saturated steam at the temperature $t°$ C. is $P = \left(\dfrac{40 + t}{140}\right)^5$ as long as t is above 80°. Find the rate of variation of the pressure with the temperature at 100° C.

Since $\qquad P = \left(\dfrac{40 + t}{140}\right)^5; \quad \dfrac{dP}{dt} = \dfrac{5(40 + t)^4}{(140)^5}$

so that when $t = 100$,

$$\frac{dP}{dt} = \frac{5 \times (140)^4}{(140)^5} = \frac{5}{140} = \frac{1}{28} = 0.036$$

Thus, the rate of variation of the pressure is, when $t = 100$, 0.036 atmosphere per degree centigrade change of temperature.

Rules for Finding the Derivatives

A) If f is a constant function, $f(x) = c$, then $f'(x) = 0$.

B) If $\boxed{f'(x) = x, \text{ then } f'(x) = 1.}$

C) If f is differentiable, then $\boxed{(cf(x))' = cf'(x)}$

D) **Power Rule**   If $f(x) = x^n$, n€ Z, then

$f'(x) = nx^{n-1}$; if $n < 0$ then x^n is not defined at $x = 0$.

E) If f and g are differentiable on the interval (a,b) then:

a) $$(f+g)'(x) = f'(x) + g'(x)$$

b) **Product Rule.** $$(fg)'(x) = f(x)g'(x) + g(x)f'(x)$$

Example: Find $f'(x)$ if $f(x) = (x^3+1)(2x^2+8x-5)$.

$f'(x) = (x^3+1)(4x+8)+(2x^2+8x-5)(3x^2)$

$\qquad = 4x^4 + 8x^3 + 4x + 8 + 6x^4 + 24x^3 - 15x^2$

$\qquad = 10x^4 + 32x^3 - 15x^2 + 4x + 8$

c) **Quotient Rule:** $$\left(\frac{f'}{g}\right)(x) = \frac{g(x)f'(x) - f(x)g'(x)}{[g(x)]^2}$$

Example: Find $f'(x)$ if $f(x) = \frac{3x^2-x+2}{4x^2+5}$

$f'(x) = \frac{-(3x^2-x+2)(8x)+(4x^2+5)(6x-1)}{(4x^2+5)^2}$

$\qquad = \frac{-(24x^3-8x^2+16x)+(24x^3-4x^2+30x-5)}{(4x^2+5)^2}$

$\qquad = \frac{4x^2+14x-5}{(4x^2+5)^2}$

F) If $f(x) = x^{m/n}$, then $f'(x) = \frac{m}{n} x^{\frac{m}{n}-1}$
where m ,n € Z and n ≠ 0

G) Polynomials. If $f(x) = (a_0+a_1x+a_2x^2+\ldots+a_nx^n)$
then $f'(x) = a_1+2a_2x+3a_3x^2+\ldots+na_nx^{n-1}$
This employs the power rule and rules concerning constants.

45

H) Chain Rule. Let f(u) be a composite function, where u=g(x).

Then f'(u) = f'(u)g'(x) or if y=f(u) and u=g(x) then $D_x y = (D_u y)(D_x u) = f'(u)g'(x)$

Additional Problem Solving Examples

Find the derivative of: $y = x^{3b}$.

Applying the theorem for $d(u^n)$,

$$\frac{dy}{dx} = 3b \cdot x^{3b-1} .$$

Find the derivative of: $y = (x^2 + 2)^3$.

Method 1. We may expand the cube and write:

$$\frac{dy}{dx} = \frac{d}{dx} [(x^2 + 2)^3] = \frac{d}{dx} (x^6 + 6x^4 + 12x^2 + 8)$$

$$= 6x^5 + 24x^3 + 24x.$$

Method 2. Let $u = x^2 + 2$, then $y = (x^2 + 2)^3 = u^3$;

Using the chain rule we have:

$$\frac{dy}{dx} = \frac{dy}{du} \cdot \frac{du}{dx} = \frac{d(u^3)}{du} \cdot \frac{d(x^2 + 2)}{dx} = 3u^2(2x)$$

$$= 3(x^2 + 2)^2 \cdot (2x) = 3(x^4 + 4x^2 + 4) \cdot (2x)$$

$$= 6x^5 + 24x^3 + 24x.$$

Exercises III

(See answers on page 297)

(1) Differentiate

(a) $u = 1 + x + \dfrac{x^2}{1 \times 2} + \dfrac{x^3}{1 \times 2 \times 3} + \ldots$

(b) $y = ax^2 + bx + c$ (c) $y = (x + a)^2$

(d) $y = (x + a)^3$

(2) If $w = at - \frac{1}{2}bt^2$, find $\dfrac{aw}{dt}$

(3) Find the derivative of

$$y = \left(x + \sqrt{-1}\right) \times \left(x - \sqrt{-1}\right)$$

(4) Differentiate

$$y = (197x - 34x^2) \times (7 + 22x - 83x^3)$$

(5) If $x = (y + 3) \times (y + 5)$, find $\dfrac{dx}{dy}$.

(6) Differentiate $y = 1.3709x \times (112.6 + 45.202x^2)$.

Find the derivatives of

(7) $y = \dfrac{2x + 3}{3x + 2}$ (8) $y = \dfrac{1 + x + 2x^2 + 3x^3}{1 + x + 2x^2}$

(9) $y = \dfrac{ax + b}{cx + d}$ (10) $y = \dfrac{x^n + a}{x^{-n} + b}$

(11) The temperature t of the filament of an incandescent electric lamp is connected to the current passing through the lamp by the relation

$$C = a + bt + ct^2$$

Find an expression giving the variation of the current corresponding to a variation of temperature.

(12) The following formulae have been proposed to express the relation between the electric resistance R of a wire at the temperature $t°$ C., and the resistance R_0 of that same wire at $0°$ centigrade, a and b being constants.

47

$$R = R_0(1 + at + bt^2)$$

$$R = R_0\left(1 + at + b\sqrt{t}\right)$$

$$R = R_0(1 + at + bt^2)^{-1}$$

Find the rate of variation of the resistance with regard to temperature as given by each of these formulae.

(13) The electromotive force E of a certain type of standard cell has been found to vary with the temperature t according to the relation

$$E = 1.4340[1 - 0.000814(t - 15) + 0.000007(t - 15)^2] \text{ volts}$$

Find the change of electromotive force per degree, at $15°$, $20°$ and $25°$.

(14) The electromotive force necessary to maintain an electric arc of length l with a current intensity i has been found to be

$$E = a + bl + \frac{c + kl}{i}$$

where a, b, c, k are constants.

Find an expression for the variation of the electromotive force (*a*) with regard to the length of the arc; (*b*) with regard to the strength of the current.

CHAPTER 7

Successive Differentiation

When we are given a function

$$s = f(t)$$

where s is distance and t is time, the first derivative of $f(t) = f'(t)$ gives the speed or the rate at which the position is changing. Such a situation can be related to a moving car, for example.

The derivative of $f'(t) = f''(t)$ gives the acceleration or the rate at which the speed is changing. $f''(t)$ is referred to as the second derivative of $f(t)$.

The derivative of $f''(t) = f'''(t)$ is the third derivative of $f(t)$. $f'''(t)$ gives the rate at which the acceleration is changing.

In solving practical problems, in our universe as we know it, there is seldom a need to go to derivatives that are higher than the third. Higher derivatives, however, can be involved in abstract mathematical operations.

Let us try the effect of repeating several times over the operation of differentiating a function. Begin with a concrete case.

Let $y = x^5$.

First differentiation, $5x^4$.
Second differentiation, $5 \times 4x^3$ $= 20x^3$.
Third differentiation, $5 \times 4 \times 3x^2$ $= 60x^2$.
Fourth differentiation, $5 \times 4 \times 3 \times 2x$ $= 120x$.
Fifth differentiation, $5 \times 4 \times 3 \times 2 \times 1 = 120$.
Sixth differentiation, $= 0$.

49

There is a certain notation, with which we are already acquainted, used by some writers, that is very convenient. This is to employ the general symbol $f(x)$ for any function of x. Here the symbol $f(\ )$ is read as "function of", without saying what particular function is meant. So the statement $y = f(x)$ merely tells us that y is a function of x, it may be x^2 or ax^n, or $\cos x$ or any other complicated function of x.

The corresponding symbol for the derivative is $f'(x)$, which is simpler to write than $\dfrac{dy}{dx}$. This is called the "derived function" of x.

Suppose we differentiate over again, we shall get the "second derived function" or second derivative which is denoted by $f''(x)$; and so on.

Now let us generalize.

Let $y = f(x) = x^n$.

First differentiation, $f'(x) = nx^{n-1}$.
Second differentiation, $f''(x) = n(n-1)x^{n-2}$.
Third differentiation, $f'''(x) = n(n-1)(n-2)x^{n-3}$.
Fourth differentiation, $f''''(x) = n(n-1)(n-2)(n-3)x^{n-4}$.

           etc., etc.

But this is not the only way of indicating successive differentiations. For, if the original function be

$$y = f(x);$$

differentiating once gives $\dfrac{dy}{dx} = f'(x);$

differentiating twice gives $\dfrac{d\left(\dfrac{dy}{dx}\right)}{dx} = f''(x);$

50

and this is more conveniently written as $\dfrac{d^2y}{(dx)^2}$, or more usually

$\dfrac{d^2y}{dx^2}$. Similarly, we may write as the result of differentiating three

times, $\dfrac{d^3y}{dx^3} = f'''(x)$.

Examples.

Now let us try $\quad y = f(x) = 7x^4 + 3.5x^3 - \tfrac{1}{2}x^2 + x - 2$

$$\frac{dy}{dx} = f'(x) \quad = 28x^3 + 10.5x^2 - x + 1$$

$$\frac{d^2y}{dx^2} = f''(x) \quad = 84x^2 + 21x - 1$$

$$\frac{d^3y}{dx^3} = f'''(x) \quad = 168x + 21$$

$$\frac{d^4y}{dx^4} = f''''(x) \ = 168$$

$$\frac{d^5y}{dx^5} = f'''''(x) = 0$$

In a similar manner if $\quad y = \phi(x) = 3x(x^2 - 4)$

$$\phi'(x) = \frac{dy}{dx} \ = 3[x \times 2x + (x^2 - 4) \times 1] = 3(3x^2 - 4)$$

$$\phi''(x) = \frac{d^2y}{dx^2} = 3 \times 6x = 18x$$

51

$$\phi'''(x) = \frac{d^3y}{dx^3} = 18$$

$$\phi''''(x) = \frac{d^4y}{dx^4} = 0$$

Examples.

Find the sixth derivative of $y = x^6$.

First derivative $= 6x^{6-1} = 6x^5$

Second derivative $= 5 \cdot 6x^{5-1} = 30x^4$

Third derivative $= 4 \cdot 30x^{4-1} = 120x^3$

Fourth derivative $= 3 \cdot 120x^{3-1} = 360x^2$

Fifth derivative $= 2 \cdot 360x^{2-1} = 720x^1 = 720x$

Sixth derivative $= 1 \cdot 720x^{1-1} = 720x^0 = 720$

The seventh derivative is seen to be zero, and therefore the function $y = x^6$ has seven derivatives.

Find $y'' \dfrac{d^2y}{dx^2}$ for the expression $xy^3 = 1$.

To find the second derivative, y", we must first find the first derivative and then differentiate that to obtain the second derivative.

We could solve for y and then differentiate to obtain y', but an alternative is implicit differentiation,

$$xy^3 = 1.$$

Differentiating implicitly,

$$3xy^2 \cdot y' + y^3 = 0.$$

$$3xy^2 \cdot y' = -y^3.$$

$$y' = \frac{-y^3}{3xy^2}$$

$$= -\frac{y}{3x}.$$

Now we take the derivative of y' to find y".

$$y'' = -\frac{1}{3}\left[\frac{x \cdot y' - y}{x^2}\right].$$

Substituting $y' = -\dfrac{y}{3x}$ in the expression for y" and simplifying

$$y'' = -\frac{1}{3}\left[\frac{-x\left(\dfrac{-y}{3x}\right) - y}{x^2}\right]$$

$$= -\frac{1}{3}\left[\frac{-\dfrac{y}{3} - y}{x^2}\right]$$

$$= -\frac{1}{3} \left[\frac{-\frac{4}{3}y}{x^2} \right]$$

$$= \frac{4y}{9x^2} \; .$$

Exercises IV

(See answers on page 297)

Find $\dfrac{dy}{dx}$ and $\dfrac{d^2y}{dx^2}$ for the following expressions:

(1) $y = 17x + 12x^2$

(2) $y = \dfrac{x^2 + a}{x + a}$

(3) $y = 1 + \dfrac{x}{1} + \dfrac{x^2}{1 \times 2} + \dfrac{x^3}{1 \times 2 \times 3} + \dfrac{x^4}{1 \times 2 \times 3 \times 4}$

(4) Find the 2nd and 3rd derivatives in Exercises III, No. 1 to No. 7, and in the examples given in Chapter 6, No. 1 to No. 7.

CHAPTER 8

When Time Varies

Application Of Time And Motion

Some of the most important problems of the calculus are those where time is the independent variable, and we have to think about the values of some other quantity that varies when the time varies. Some things grow larger as time goes on; some other things grow smaller. The distance that a train has travelled from its starting place goes on ever increasing as time goes on. Trees grow taller as the years go by. Which is growing at the greater rate: a plant 12 inches high which in one month becomes 14 inches high, or a tree 12 feet high which in a year becomes 14 feet high?

In this chapter we are going to make much use of the word *rate*. Nothing to do with birth rate or death rate, though these words suggest so many births or deaths per thousand of the population. When a car whizzes by us, we say: What a terrific rate! When a spendthrift is flinging about his money, we remark that that young man is living at a prodigious rate. What do we mean by *rate*? In both these cases we are making a mental comparison of something that is happening, and the length of time it takes to happen. If the car goes 10 yards per second, a simple bit of mental arithmetic will show us that this is equivalent—while it lasts—to a rate of 600 yards per minute, or over 20 miles per hour.

Now in what sense is it true that a speed of 10 yards per second is the same as 600 yards per minute? Ten yards is not the same as 600 yards, nor is one second the same thing as one minute. What we mean by saying that the *rate* is the same, is this: that the proportion borne between distance passed over and time taken to pass over it, is the same in both cases.

Now try to put some of these ideas into differential notation. Let y in this case stand for money, and let t stand for time. If you are spending money, and the amount you spend in a short time dt be called dy, the *rate* of spending it will be $\dfrac{dy}{dt}$; or, as regards saving, with a minus sign, as $-\dfrac{dy}{dt}$, because then dy is a *decrement*, not an increment. But money is not a good example for the calculus, because it generally comes and goes by jumps, not by a continuous flow—you may earn \$20,000 a year, but it does not keep running in all day long in a thin stream; it comes in only weekly, or monthly, or quarterly, in lumps: and your expenditure also goes out in sudden payments.

A more apt illustration of the idea of a rate is furnished by the speed of a moving body. From London to Liverpool is 200 miles. If a train leaves London at 7 o'clock, and reaches Liverpool at 11 o'clock, you know that, since it has travelled 200 miles in 4 hours, its average rate must have been 50 miles per hour; because $\frac{200}{4} = \frac{50}{1}$. Here you are really making a mental comparison between the distance passed over and the time taken to pass over it. You are dividing one by the other. If y is the whole distance, and t the whole time, clearly the average rate is $\dfrac{y}{t}$. Now the speed was not actually constant all the way: at starting, and during the slowing up at the end of the journey, the speed was less. Probably at some part, when running downhill, the speed was over 60 miles an hour. If, during any particular element of time dt, the corresponding element of distance passed over was dy, then at that part of the journey the speed was $\dfrac{dy}{dt}$. The *rate* at which one quantity (in the present instance, *distance*) is changing in relation to the other quantity (in this case, *time*) is properly expressed, then, by stating the derivative of one with respect to the other. A *velocity*, scientifically expressed, is the rate at which a very small distance in any given direction is being passed over, and may therefore be written

$$v = \frac{dy}{dt}$$

But if the velocity v is not uniform, then it must be either increasing or else decreasing. The rate at which a velocity is increasing is called the *acceleration*. If a moving body is, at any particular instant, gaining an additional velocity dv in an element of time dt, then the acceleration a at that instant may be written

$$a = \frac{dv}{dt}$$

But since $v = \frac{dy}{dt}$,

$$a = \frac{dv}{dt} = \frac{d}{dt}\left(\frac{dy}{dt}\right)$$

which is usually written $\quad a = \frac{d^2y}{dt^2}$;

or the acceleration is the second derivative of the distance, with respect to time. Acceleration is expressed as a change of velocity in unit time, for instance, as being so many feet per second per second; the notation used being ft/sec^2.

When a railway train has just begun to move, its velocity v is small; but it is rapidly gaining speed—it is being hurried up, or accelerated, by the effort of the engine. So its $\frac{d^2y}{dt^2}$ is large. When it has got up its top speed it is no longer being accelerated, so that then $\frac{d^2y}{dt^2}$ has fallen to zero. But when it nears its stopping place its speed begins to slow down; may, indeed, slow down very quickly if the brakes are put on, and during this period of *deceleration* or slackening of pace, the value of $\frac{dv}{dt}$, that is, of $\frac{d^2y}{dt^2}$, will be negative.

To accelerate a mass m requires the continuous application of force. The force necessary to accelerate a mass is proportional to

57

the mass, and it is also proportional to the acceleration which is being imparted. Hence we may write for the force f, the expression

$$f = ma$$

or

$$f = m\frac{dv}{dt}$$

or

$$f = m\frac{d^2y}{dt^2}$$

The product of a mass by the speed at which it is going is called its *momentum*, and is in symbols mv. If we differentiate momentum with respect to time we shall get $\dfrac{d(mv)}{dt}$ for the rate of change of momentum. But, since m is a constant quantity, this may be written $m\dfrac{dv}{dt}$, which we see above is the same as f. That is to say, force may be expressed either as mass times acceleration, or as rate of change of momentum.

Again, if a force is employed to move something (against an equal and opposite counter-force), it does *work*; and the amount of work done is measured by the product of the force into the distance (in its own direction) through which its point of application moves forward. So if a force f moves forward through a length y, the work done (which we may call w) will be

$$w = f \times y$$

where we take f as a constant force. If the force varies at different parts of the range y, then we must find an expression for its value from point to point. If f be the force along the small element of length dy, the amount of work done will be $f \times dy$. But as dy is only an element of length, only an element of work will be done. If we write w for work, then an element of work will be dw; and we have

$$dw = f \times dy$$

which may be written $\quad dw = ma \cdot dy;$

or

$$dw = m\frac{d^2y}{dt^2} \cdot dy$$

58

or
$$dw = m\frac{dv}{dt} \cdot dy$$

Further, we may transpose the expression and write

$$\frac{dw}{dy} = f$$

This gives us yet a third definition of *force;* that if it is being used to produce a displacement in any direction, the force (in that direction) is equal to the rate at which work is being done per unit of length in that direction. In this last sentence the word *rate* is clearly not used in its time-sense, but in its meaning as ratio or proportion.

Sir Isaac Newton, who was (along with Leibniz) an inventor of the methods of the calculus, regarded all quantities that were varying as *flowing;* and the ratio which we nowadays call the derivative he regarded as the rate of flowing, or the *fluxion* of the quantity in question. He did not use the notation of the dy and dx, and dt (this was due to Leibniz), but had instead a notation of his own. If y was a quantity that varied, or "flowed", then his symbol for its rate of variation (or "fluxion") was $\dot{y}$. If x was the variable, then its fluxion was called $\dot{x}$. The dot over the letter indicated that it had been differentiated. But this notation does not tell us what is the independent variable with respect to which the differentiation has been effected. When we see $\frac{dy}{dt}$ we know that y is to be differentiated with respect to t. If we see $\frac{dy}{dx}$ we know that y is to be differentiated with respect to x. But if we see merely $\dot{y}$, we cannot tell without looking at the context whether this is to mean $\frac{dy}{dx}$ or $\frac{dy}{dt}$ or $\frac{dy}{dz}$, or what is the other variable. So, therefore, this fluxional notation is less informing than the differential notation, and has in consequence largely dropped out of use. But its simplicity gives it an advantage if only we will agree to use it for those cases exclusively where *time* is the independent variable. In that case $\dot{y}$ will mean $\dfrac{dy}{dt}$ and $\dot{u}$ will mean $\dfrac{du}{dt}$; and $\ddot{x}$ will mean $\dfrac{d^2x}{dt^2}$.

Adopting this fluxional notation we may write the mechanical equations considered in the paragraphs above, as follows:

distance	x
velocity	$v = \dot{x}$
acceleration	$a = \dot{v} = \ddot{x}$
force	$f = m\dot{v} = m\ddot{x}$
work	$w = x \times m\ddot{x}$

Examples.
(1) A body moves so that the distance x (in feet), which it travels from a certain point O, is given by the relation

$$x = 0.2t^2 + 10.4$$

where t is the time in seconds elapsed since a certain instant. Find the velocity and acceleration 5 seconds after the body began to move, and also find the corresponding values when the distance covered is 100 feet. Find also the average velocity during the first 10 seconds of its motion. (Suppose distances and motion to the right to be positive.)

Now $\qquad\qquad x = 0.2t^2 + 10.4$

$$v = \dot{x} = \frac{dx}{dt} = 0.4t; \quad \text{and} \quad a = \ddot{x} = \frac{d^2x}{dt^2} = 0.4 = \text{constant.}$$

When $t = 0$, $x = 10.4$ and $v = 0$. The body started from a point 10.4 feet to the right of the point O; and the time was reckoned from the instant the body started.

When $t = 5$, $v = 0.4 \times 5 = 2$ ft./sec.; $a = 0.4$ ft./sec^2.

When $x = 100$, $100 = 0.2t^2 + 10.4$, or $t^2 = 448$,

and $t = 21.17$ sec.; $v = 0.4 \times 21.17 = 8.468$ ft./sec.

When $t = 10$,

distance travelled $= 0.2 \times 10^2 + 10.4 - 10.4 = 20$ ft.

Average velocity $= \frac{20}{10} = 2$ ft./sec.

(It is the same velocity as the velocity at the middle of the interval, $t = 5$; for, the acceleration being constant, the velocity has varied uniformly from zero when $t = 0$ to 4 ft./sec. when $t = 10$.)

(2) In the above problem let us suppose

$$x = 0.2t^2 + 3t + 10.4$$

$$v = \dot{x} = \frac{dx}{dt} = 0.4t + 3; \quad a = \ddot{x} = \frac{d^2x}{dt^2} = 0.4 = \text{constant}.$$

When $t = 0$, $x = 10.4$ and $v = 3$ ft./sec., the time is reckoned from the instant at which the body passed a point 10.4 ft. from the point O, its velocity being then already 3 ft./sec. To find the time elapsed since it began moving, let $v = 0$; then $0.4t + 3 = 0$, $t = -\frac{3}{4} = -7.5$ sec. The body began moving 7.5 sec. before time was begun to be observed; 5 seconds after this gives $t = -2.5$ and $v = 0.4 \times -2.5 + 3 = 2$ ft./sec.

When $x = 100$ ft.,

$$100 = 0.2t^2 + 3t + 10.4; \quad \text{or} \quad t^2 + 15t - 448 = 0$$

hence $\quad t = 14.96$ sec., $v = 0.4 \times 14.96 + 3 = 8.98$ ft./sec.

To find the distance travelled during the first 10 seconds of the motion one must know how far the body was from the point O when it started.

When $t = -7.5$,

$$x = 0.2 \times (-7.5)^2 - 3 \times 7.5 + 10.4 = -0.85 \text{ ft.}$$

that is 0.85 ft. to the left of the point O.

Now, when $t = 2.5$,

$$x = 0.2 \times 2.5^2 + 3 \times 2.5 + 10.4 = 19.15$$

So, in 10 seconds, the distance travelled was $19.15 + 0.85 = 20$ ft., and

the average velocity $= \frac{20}{10} = 2$ ft./sec.

(3) Consider a similar problem when the distance is given by $x = 0.2t^2 - 3t + 10.4$. Then $v = 0.4t - 3$, $a = 0.4 =$ constant. When $t = 0$, $x = 10.4$ as before, and $v = -3$; so that the body was moving in the direction opposite to its motion in the previous cases. As the acceleration is positive, however, we see that this velocity will decrease as time goes on, until it becomes zero, when $v = 0$ or $0.4t - 3 = 0$; or $t = +7.5$ sec. After this, the velocity becomes positive; and 5 seconds after the body started, $t = 12.5$, and

$$v = 0.4 \times 12.5 - 3 = 2 \text{ ft./sec.}$$

When $x = 100$,

$$100 = 0.2t^2 - 3t + 10.4, \quad \text{or} \quad t^2 - 15t - 448 = 0$$

and $\quad t = 29.96$; $v = 0.4 \times 29.96 - 3 = 8.98$ ft./sec.

When v is zero, $x = 0.2 \times 7.5^2 - 3 \times 7.5 + 10.4 = -0.85$, informing us that the body moves back to 0.85 ft. beyond the point O before it stops. Ten seconds later $t = 17.5$ and

$$x = 0.2 \times 17.5^2 - 3 \times 17.5 + 10.4 = 19.15$$

The distance travelled $= 0.85 + 19.15 = 20.0$, and the average velocity is again 2 ft./sec.

(4) Consider yet another problem of the same sort with $x = 0.2t^3 - 3t^2 + 10.4$; $v = 0.6t^2 - 6t$; $a = 1.2t - 6$. The acceleration is no more constant.

When $t = 0$, $x = 10.4$, $v = 0$, $a = -6$. The body is at rest, but just ready to move with a negative acceleration, that is, to gain a velocity towards the point O.

(5) If we have $x = 0.2t^3 - 3t + 10.4$, then $v = 0.6t^2 - 3$, and $a = 1.2t$.

When $t = 0$, $x = 10.4$; $v = -3$; $a = 0$.

The body is moving towards the point O with a velocity of 3 ft./sec., and just at that instant the velocity is uniform.

We see that the conditions of the motion can always be at once ascertained from the time-distance equation and its first and second derived functions. In the last two cases the mean velocity during the first 10 seconds and the velocity 5 seconds after the start will no more be the same, because the velocity is not increasing uniformly, the acceleration being no longer constant.

(6) The angle θ (in radians) turned through by a wheel is given by $\theta = 3 + 2t - 0.1t^3$, where t is the time in seconds from a certain instant; find the angular velocity ω and the angular acceleration α, (*a*) after 1 second; (*b*) after it has performed one revolution. At what time is it at rest, and how many revolutions has it performed up to that instant?

$$\omega = \dot{\theta} = \frac{d\theta}{dt} = 2 - 0.3t^2, \quad \alpha = \ddot{\theta} = \frac{d^2\theta}{dt^2} = -0.6t$$

When $t = 0$, $\theta = 3$; $\omega = 2$ rad./sec.; $\alpha = 0$.
When $t = 1$, $\omega = 2 - 0.3 = 1.7$ rad./sec.; $\alpha = -0.6$ rad./sec^2.
This is a retardation; the wheel is slowing down.
After 1 revolution

$$\theta = 2\pi = 3 + 2t - 0.1t^3$$

By solving this equation numerically we can get the value or values of t for which $\theta = 2\pi$; these are about 2.11 and 3.02 (there is a third negative value).

When $t = 2.11$,

$$\theta = 6.28; \quad \omega = 2 - 1.34 = 0.66 \text{ rad./sec.}$$

$$\alpha = -1.27 \text{ rad./sec}^2$$

When $t = 3.02$,

$$\theta = 6.28; \quad \omega = 2 - 2.74 = -0.74 \text{ rad./sec.}$$

$$\alpha = -1.81 \text{ rad./sec}^2$$

The velocity is reversed. The wheel is evidently at rest between these two instants; it is at rest when $\omega = 0$, that is when $0 = 2 - 0.3t^2$, or when $t = 2.58$ sec., it has performed

$$\frac{\theta}{2\pi} = \frac{3 + 2 \times 2.58 - 0.1 \times 2.58^3}{6.28} = 1.025 \text{ revolutions}$$

Additional Problem Solving Examples

Q A rope attached to a boat is being pulled in at a rate of 10 ft/sec. If the water is 20 ft below the level at which the rope is being drawn in, how fast is the boat approaching the wharf when 36 ft of rope are yet to be pulled in?

A

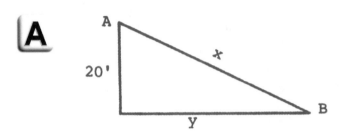

The length AB denotes the rope, and the position of the boat is at B. Since the rope is being drawn in at a rate of 10ft/sec.

$$\frac{dx}{dt} = 10.$$

To find how fast the boat is being towed in when 36 ft. of rope are left,

$$\frac{dy}{dt} \text{ must be found at } x = 36.$$

From the right triangle, $20^2 + y^2 = x^2$ or $y = \sqrt{x^2 - 400}$. Differentiating with respect to t,

$$\frac{dy}{dt} = \frac{dy}{dx} \cdot \frac{dx}{dt} = \frac{1}{2} (x^2 - 400)^{-\frac{1}{2}} (2x) \frac{dx}{dt} = \frac{x \frac{dx}{dt}}{\sqrt{x^2 - 400}}.$$

Substituting the conditions that:

$$\frac{dx}{dt} = -10, \text{ and } x = 36,$$

$$\frac{dy}{dt} = \frac{-360}{\sqrt{896}} = -\frac{45}{\sqrt{14}}.$$

It has now been found that, when there are 36 ft of rope left, the boat is moving in at the rate of:

$$\frac{45}{\sqrt{14}} \text{ ft/sec.}$$

 A boat is being hauled toward a pier at a height of 20 ft above the water level. The rope is drawn in at a rate of 6ft/ sec. Neglecting sag, how fast is the boat approaching the base of the pier when 25 ft of rope remain to be pulled in?

A

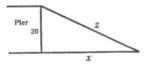

Formulating the given data, we have:

$$\frac{dz}{dt} = 6, z = 25, \text{ and} \frac{dx}{dt} \text{ is to be found.}$$

At any time t we have, from the Pythagorean theorem,

$$20^2 + x^2 = z^2$$

By differentiation, we obtain:

$$x\frac{dx}{dt} = z\frac{dz}{dt}$$

When $z = 25$, $x = \sqrt{25^2 - 20^2} = 15$; therefore

$$15 \frac{dx}{dt} = 25\,(-6)$$

$$\frac{dx}{dt} = -10 \text{ ft/sec}$$

(The boat approaches the base at 10 ft/sec).

Q Compute the average rate of change of $y = f(x) = x^2 - 2$ between $x = 3$ and $x = 4$.

A Average rate of change is defined as:

$\dfrac{\Delta y}{\Delta x}$ with $\Delta y = f(x + \Delta x) - f(x)$.

Given: $x = 3$, $\Delta x = 4 - 3 = 1$,

$y = f(x) = f(3) = 3^2 - 2 = 7$

For $x = 4$,

$y + \Delta y = f(x + \Delta x) = 4^2 - 2 = 14$

$\Delta y = f(x + \Delta x) - f(x) = f(4) - f(3)$

$\qquad = (4^2 - 2) - (3^2 - 2) = 14 - 7 = 7$

$\dfrac{\Delta y}{\Delta x} = \dfrac{7}{1} = 7$, the average rate of change.

Q Find the rate of change of y with respect to x at the point $x = 5$, if $2y = x^2\ 3x - 1$.

A Rate of change is defined as

$\displaystyle\lim_{\Delta x \to 0} \dfrac{\Delta y}{\Delta x}$, with

$\Delta y = f(x + \Delta x) - f(x)$.

We have:

$$2\Delta y = (x + \Delta x)^2 + 3(x + \Delta x) - 1 - (x^2 + 3x - 1)$$

$$= x^2 + 2x \cdot \Delta x + (\Delta x)^2 + 3x + 3\Delta x - 1 - x^2 - 3x + 1$$

$$= 2x \cdot \Delta x + (\Delta x)^2 + 3\Delta x.$$

Dividing by Δx,

$$\frac{2\,\Delta y}{\Delta x} = \frac{2x \cdot \Delta x}{\Delta x} + \frac{(\Delta x)^2}{\Delta x} + \frac{3\,\Delta x}{\Delta x}$$

$$= 2x + \Delta x + 3$$

and

$$\frac{\Delta y}{\Delta x} = x + \frac{\Delta x}{2} + \frac{3}{2}\ .$$

Now,

$$\lim_{\Delta x \to 0} \frac{\Delta y}{\Delta x} = \lim_{\Delta x \to 0} x + \frac{\Delta x}{2} + \frac{3}{2} = x + \frac{3}{2}.$$

For $x = 5$,

$$\lim_{\Delta x \to 0} \frac{\Delta y}{\Delta x} = 5 + \frac{3}{2} = 6\frac{1}{2}$$

This means that the instantaneous rate of change of the function represented by the curve at the point $x = 5$ is $6\frac{1}{2}$.

The function, it is seen, changes $6\frac{1}{2}$ times as fast as the independent variable x at $x = 5$.

The slope of the tangent at $x = 5$ is $6\frac{1}{2}$.

Exercises V

(See answers on page 298)

(1) If $y = a + bt^2 + ct^4$; find $\dfrac{dy}{dt}$ and $\dfrac{d^2y}{dt^2}$.

Ans. $\dfrac{dy}{dt} = 2bt + 4ct^3$; $\dfrac{d^2y}{dt^2} = 2b + 12ct^2$

(2) A body falling freely in space describes in t seconds a space s, in feet, expressed by the equation $s = 16t^2$. Draw a curve showing the relation between s and t. Also determine the velocity of the body at the following times from its being let drop: $t = 2$ seconds; $t = 4.6$ seconds; $t = 0.01$ second.

(3) If $x = at - \frac{1}{2}gt^2$; find $\dot{x}$ and $\ddot{x}$.

(4) If a body moves according to the law

$$s = 12 - 4.5t + 6.2t^2$$

find its velocity when $t = 4$ seconds; s being in feet.

(5) Find the acceleration of the body mentioned in the preceding example. Is the acceleration the same for all values of t?

(6) The angle θ (in radians) turned through by a revolving wheel is connected with the time t (in seconds) that has elapsed since starting, by the law

$$\theta = 2.1 - 3.2t + 4.8t^2$$

Find the angular velocity (in radians per second) of that wheel when $1\frac{1}{2}$ seconds have elapsed. Find also its angular acceleration.

(7) A slider moves so that, during the first part of its motion, its distance s in inches from its starting point is given by the expression

$$s = 6.8t^3 - 10.8t; \ t \text{ being in seconds.}$$

Find the expression for the velocity and the acceleration at any time; and hence find the velocity and the acceleration after 3 seconds.

(8) The motion of a rising balloon is such that its height h, in miles, is given at any instant by the expression

68

$h = 0.5 + \frac{1}{10}\sqrt[3]{t - 125}$; t being in seconds.

Find an expression for the velocity and the acceleration at any time. Draw curves to show the variation of height, velocity and acceleration during the first ten minutes of the ascent.

(9) A stone is thrown downwards into water and its depth p in meters at any instant t seconds after reaching the surface of the water is given by the expression

$$p = \frac{4}{4 + t^2} + 0.8t - 1$$

Find an expression for the velocity and the acceleration at any time. Find the velocity and acceleration after 10 seconds.

(10) A body moves in such a way that the space described in the time t from starting is given by $s = t^n$, where n is a constant. Find the value of n when the velocity is doubled from the 5th to the 10th second; find it also when the velocity is numerically equal to the acceleration at the end of the 10th second.

CHAPTER 9

Introducing A Useful Dodge
—The Chain Rule

There are complex expressions which are difficult to differentiate directly. For this purpose, the chain rule often helps to convert the complex expression into one that is simpler and can be differentiated by the basic rules. With the chain rule we compute a series or chain of derivatives and simply multiply them together.

In the preceding chapters on differentiation we dealt with expressions that could be differentiated directly using the basic rules. In this chapter we deal with expressions that need to be operated on at first before the basic rules of differentiation can be applied.

Sometimes one is stumped by finding that the expression to be differentiated is too complicated to tackle directly.

Thus, the equation

$$y = (x^2 + a^2)^{\frac{3}{2}}$$

is awkward to a beginner.

Now the dodge to turn the difficulty is this: Write some symbol, such as u, for the expression $x^2 + a^2$; then the equation becomes

$$y = u^{\frac{3}{2}}$$

which you can easily manage; for

$$\frac{dy}{du} = \frac{3}{2}u^{\frac{1}{2}}$$

Then tackle the expression

$$u = x^2 + a^2$$

and differentiate it with respect to x

$$\frac{du}{dx} = 2x$$

Then all that remains is plain sailing;

for

$$\frac{dy}{dx} = \frac{dy}{du} \times \frac{du}{dx}$$

that is,

$$\frac{dy}{dx} = \frac{3}{2}u^{\frac{1}{2}} \times 2x$$

$$= \frac{3}{2}(x^2 + a^2)^{\frac{1}{2}} \times 2x$$

$$= 3x(x^2 + a^2)^{\frac{1}{2}}$$

and so the trick is done.

By and by, when you have learned how to deal with sines, and cosines, and exponentials, you will find this dodge of increasing usefulness.

Examples.

Let us practice this dodge on a few examples.

(1) Differentiate $y = \sqrt{a + x}$.

Let $u = a + x$.

$$\frac{du}{dx} = 1; \ y = u^{\frac{1}{2}}; \ \frac{dy}{du} = \frac{1}{2}u^{-\frac{1}{2}} = \frac{1}{2}(a + x)^{-\frac{1}{2}}$$

$$\frac{dy}{dx} = \frac{dy}{du} \times \frac{du}{dx} = \frac{1}{2\sqrt{a + x}}$$

(2) Differentiate $y = \dfrac{1}{\sqrt{a + x^2}}$

Let $u = a + x^2$.

$$\frac{du}{dx} = 2x; \quad y = u^{-\frac{1}{2}}; \quad \frac{dy}{du} = -\frac{1}{2}u^{-\frac{3}{2}}$$

$$\frac{dy}{dx} = \frac{dy}{du} \times \frac{du}{dx} = -\frac{x}{\sqrt{(a + x^2)^3}}$$

(3) Differentiate $y = \left(m - nx^{\frac{2}{3}} + \dfrac{p}{x^{\frac{4}{3}}} \right)^{a}$

Let $u = m - nx^{\frac{2}{3}} + px^{-\frac{4}{3}}$.

$$\frac{du}{dx} = -\frac{2}{3}nx^{-\frac{1}{3}} - \frac{4}{3}px^{-\frac{7}{3}}$$

$$y = u^{a}; \quad \frac{dy}{du} = au^{a-1}$$

$$\frac{dy}{dx} = \frac{dy}{du} \times \frac{du}{dx} = -a\left(m - nx^{\frac{2}{3}} + \frac{p}{x^{\frac{4}{3}}} \right)^{a-1} \left(\frac{2}{3}nx^{-\frac{1}{3}} + \frac{4}{3}px^{-\frac{7}{3}} \right)$$

(4) Differentiate $y = \dfrac{1}{\sqrt{x^3 - a^2}}$

Let $u = x^3 - a^2$.

$$\frac{du}{dx} = 3x^2; \quad y = u^{-\frac{1}{2}}; \quad \frac{dy}{du} = -\frac{1}{2}(x^3 - a^2)^{-\frac{3}{2}}$$

$$\frac{dy}{dx} = \frac{dy}{du} \times \frac{du}{dx} = -\frac{3x^2}{2\sqrt{(x^3 - a^2)^3}}$$

(5) Differentiate $y = \sqrt{\dfrac{1 - x}{1 + x}}$

Write this as $\quad y = \dfrac{(1 - x)^{\frac{1}{2}}}{(1 + x)^{\frac{1}{2}}}$

$$\frac{dy}{dx} = \frac{(1+x)^{\frac{1}{2}}\dfrac{d(1-x)^{\frac{1}{2}}}{dx} - (1-x)^{\frac{1}{2}}\dfrac{d(1+x)^{\frac{1}{2}}}{dx}}{1+x}$$

(We may also write $y = (1-x)^{\frac{1}{2}}(1+x)^{-\frac{1}{2}}$ and differentiate as a product.)

Proceeding as in Example (1) above, we get

$$\frac{d(1-x)^{\frac{1}{2}}}{dx} = -\frac{1}{2\sqrt{1-x}}; \quad \text{and} \quad \frac{d(1+x)^{\frac{1}{2}}}{dx} = \frac{1}{2\sqrt{1+x}}$$

Hence $\quad \dfrac{dy}{dx} = -\dfrac{(1+x)^{\frac{1}{2}}}{2(1+x)\sqrt{1-x}} - \dfrac{(1-x)^{\frac{1}{2}}}{2(1+x)\sqrt{1+x}}$

$$= -\frac{1}{2\sqrt{1+x}\sqrt{1-x}} - \frac{\sqrt{1-x}}{2\sqrt{(1+x)^3}}$$

or $\quad \dfrac{dy}{dx} = -\dfrac{1}{(1+x)\sqrt{1-x^2}}$

(6) Differentiate $y = \sqrt{\dfrac{x^3}{1+x^2}}$

We may write this

$$y = x^{\frac{3}{2}}(1+x^2)^{-\frac{1}{2}}$$

$$\frac{dy}{dx} = \frac{3}{2}x^{\frac{1}{2}}(1+x^2)^{-\frac{1}{2}} + x^{\frac{3}{2}} \times \frac{d\left[(1+x^2)^{-\frac{1}{2}}\right]}{dx}$$

Differentiating $(1+x^2)^{-\frac{1}{2}}$, as shown in Example (2) above, we get

$$\frac{d\left[(1+x^2)^{-\frac{1}{2}}\right]}{dx} = -\frac{x}{\sqrt{(1+x^2)^3}}$$

so that $\quad \dfrac{dy}{dx} = \dfrac{3\sqrt{x}}{2\sqrt{1+x^2}} - \dfrac{\sqrt{x^5}}{\sqrt{(1+x^2)^3}} = \dfrac{\sqrt{x}(3+x^2)}{2\sqrt{(1+x^2)^3}}$

(7) Differentiate $y = \left(x + \sqrt{x^2+x+a}\right)^3$

Let $u = x + \sqrt{x^2+x+a}$

73

$$\frac{du}{dx} = 1 + \frac{d\left[(x^2 + x + a)^{\frac{1}{2}}\right]}{dx}$$

$$y = u^3; \quad \text{and} \quad \frac{dy}{du} = 3u^2 = 3\left(x + \sqrt{x^2 + x + a}\right)^2$$

Now let $v = (x^2 + x + a)^{\frac{1}{2}}$ and $w = (x^2 + x + a)$

$$\frac{dw}{dx} = 2x + 1; \quad v = w^{\frac{1}{2}}; \quad \frac{dv}{dw} = \frac{1}{2}w^{-\frac{1}{2}}$$

$$\frac{dv}{dx} = \frac{dv}{dw} \times \frac{dw}{dx} = \frac{1}{2}(x^2 + x + a)^{-\frac{1}{2}}(2x + 1)$$

Hence $\quad \dfrac{du}{dx} = 1 + \dfrac{2x + 1}{2\sqrt{x^2 + x + a}}$

$$\frac{dy}{dx} = \frac{dy}{du} \times \frac{du}{dx}$$

$$= 3\left(x + \sqrt{x^2 + x + a}\right)^2 \left(1 + \frac{2x + 1}{2\sqrt{x^2 + x + a}}\right)$$

(8) Differentiate $y = \sqrt{\dfrac{a^2 + x^2}{a^2 - x^2}} \ \sqrt[3]{\dfrac{a^2 - x^2}{a^2 + x^2}}$

We get $\quad y = \dfrac{(a^2 + x^2)^{\frac{1}{2}}(a^2 - x^2)^{\frac{1}{3}}}{(a^2 - x^2)^{\frac{1}{2}}(a^2 + x^2)^{\frac{1}{3}}} = (a^2 + x^2)^{\frac{1}{6}}(a^2 - x^2)^{-\frac{1}{6}}$

$$\frac{dy}{dx} = (a^2 + x^2)^{\frac{1}{6}} \frac{d\left[(a^2 - x^2)^{-\frac{1}{6}}\right]}{dx} + \frac{d\left[(a^2 + x^2)^{\frac{1}{6}}\right]}{(a^2 - x^2)^{\frac{1}{6}}dx}$$

Let $u = (a^2 - x^2)^{-\frac{1}{6}}$ and $v = (a^2 - x^2)$

$$u = v^{-\frac{1}{6}}; \quad \frac{du}{dv} = -\frac{1}{6}v^{-\frac{7}{6}}; \quad \frac{dv}{dx} = -2x$$

$$\frac{du}{dx} = \frac{du}{dv} \times \frac{dv}{dx} = \frac{1}{3}x(a^2 - x^2)^{-\frac{7}{6}}$$

Let $w = (a^2 + x^2)^{\frac{1}{6}}$ and $z = (a^2 + x^2)$

$$w = z^{\frac{1}{6}};\ \frac{dw}{dz} = \frac{1}{6}z^{-\frac{5}{6}};\ \frac{dz}{dx} = 2x$$

$$\frac{dw}{dx} = \frac{dw}{dz} \times \frac{dz}{dx} = \frac{1}{3}x(a^2 + x^2)^{-\frac{5}{6}}$$

Hence

$$\frac{dy}{dx} = (a^2 + x^2)^{\frac{1}{6}}\ \frac{x}{3(a^2 - x^2)^{\frac{7}{6}}} + \frac{x}{3(a^2 - x^2)^{\frac{1}{6}}(a^2 + x^2)^{\frac{5}{6}}}$$

or

$$\frac{dy}{dx} = \frac{x}{3}\left[\sqrt[6]{\frac{a^2 + x^2}{(a^2 - x^2)^7}} + \frac{1}{\sqrt[6]{(a^2 - x^2)(a^2 + x^2)^5}}\right]$$

(9) Differentiate y^n with respect to y^5.

$$\frac{d(y^n)}{d(y^5)} = \frac{ny^{n-1}}{5y^{5-1}} = \frac{n}{5}y^{n-5}$$

(10) Find the first and second derivatives of $y = \frac{x}{b}\sqrt{(a - x)x}$.

$$\frac{dy}{dx} = \frac{x}{b}\frac{d\{[(a - x)x]^{\frac{1}{2}}\}}{dx} + \frac{\sqrt{(a - x)x}}{b}$$

Let $u = [(a - x)x]^{\frac{1}{2}}$ and let $w = (a - x)x$; then $u = w^{\frac{1}{2}}$

$$\frac{du}{dw} = \frac{1}{2}w^{-\frac{1}{2}} = \frac{1}{2w^{\frac{1}{2}}} = \frac{1}{2\sqrt{(a - x)x}}$$

$$\frac{dw}{dx} = a - 2x$$

$$\frac{du}{dw} \times \frac{dw}{dx} = \frac{du}{dx} = \frac{a - 2x}{2\sqrt{(a - x)x}}$$

Hence

$$\frac{dy}{dx} = \frac{x(a - 2x)}{2b\sqrt{(a - x)x}} + \frac{\sqrt{(a - x)x}}{b} = \frac{x(3a - 4x)}{2b\sqrt{(a - x)x}}$$

$$\text{Now} \quad \frac{d^2y}{dx^2} = \frac{2b\sqrt{(a-x)x}(3a-8x) - \dfrac{(3ax-4x^2)b(a-2x)}{\sqrt{(a-x)x}}}{4b^2(a-x)x}$$

$$= \frac{3a^2 - 12ax + 8x^2}{4b(a-x)\sqrt{(a-x)x}}$$

(We shall need these two last derivatives later on. See Ex. X, No. 11.)

Exercises VI

(See answers on page 299)

Differentiate the following:

(1) $y = \sqrt{x^2 + 1}$ (2) $y = \sqrt{x^2 + a^2}$

(3) $y = \dfrac{1}{\sqrt{a+x}}$ (4) $y = \dfrac{a}{\sqrt{a-x^2}}$ (5) $y = \dfrac{\sqrt{x^2-a^2}}{x^2}$

(6) $y = \dfrac{\sqrt[3]{x^4 + a}}{\sqrt[2]{x^3 + a}}$ (7) $y = \dfrac{a^2 + x^2}{(a+x)^2}$

(8) Differentiate y^5 with respect to y^2

(9) Differentiate $y = \dfrac{\sqrt{1 - \theta^2}}{1 - \theta}$

The process can be extended to three or more derivatives, so that $\dfrac{dy}{dx} = \dfrac{dy}{dz} \times \dfrac{dz}{dv} \times \dfrac{dv}{dx}$

Examples.

(1) If $z = 3x^4$; $v = \dfrac{7}{z^2}$; $y = \sqrt{1 + v}$, find $\dfrac{dy}{dx}$

We have $\dfrac{dy}{dv} = \dfrac{1}{2\sqrt{1+v}}$; $\dfrac{dv}{dz} = -\dfrac{14}{z^3}$; $\dfrac{dz}{dx} = 12x^3$

$$\frac{dy}{dx} = -\frac{168x^3}{\left(2\sqrt{1+v}\right)z^3} = -\frac{28}{3x^5\sqrt{9x^8 + 7}}$$

(2) If $t = \dfrac{1}{5\sqrt{\theta}}$; $x = t^3 + \dfrac{t}{2}$; $v = \dfrac{7x^2}{\sqrt[3]{x-1}}$, find $\dfrac{dv}{d\theta}$

$$\frac{dv}{dx} = \frac{7x(5x-6)}{3\sqrt[3]{(x-1)^4}}; \quad \frac{dx}{dt} = 3t^2 + \tfrac{1}{2}; \quad \frac{dt}{d\theta} = -\frac{1}{10\sqrt{\theta^3}}$$

Hence
$$\frac{dv}{d\theta} = -\frac{7x(5x-6)\left(3t^2 + \tfrac{1}{2}\right)}{30\sqrt[3]{(x-1)^4}\sqrt{\theta^3}}$$

an expression in which x must be replaced by its value, and t by its value in terms of θ.

(3) If $\theta = \dfrac{3a^2 x}{\sqrt{x^3}}$; $\omega = \dfrac{\sqrt{1-\theta^2}}{1+\theta}$ and $\phi = \sqrt{3} - \dfrac{1}{\omega\sqrt{2}}$,

find $\dfrac{d\phi}{dx}$.

We get $\theta = 3a^2 x^{-\frac{1}{2}}$; $\omega = \sqrt{\dfrac{1-\theta}{1+\theta}}$; and $\phi = \sqrt{3} - \dfrac{1}{\sqrt{2}}\omega^{-1}$

$$\frac{d\theta}{dx} = -\frac{3a^2}{2\sqrt{x^3}}; \quad \frac{d\omega}{d\theta} = -\frac{1}{(1+\theta)\sqrt{1-\theta^2}}$$

(see Example 5, Chapter 9); and

$$\frac{d\phi}{d\omega} = \frac{1}{\sqrt{2}\cdot\omega^2}$$

So that $\dfrac{d\phi}{dx} = \dfrac{1}{\sqrt{2}\times\omega^2} \times \dfrac{1}{(1+\theta)\sqrt{1-\theta^2}} \times \dfrac{3a^2}{2\sqrt{x^3}}$

Replace now first ω, then θ by its value.

Additional Problem Solving Examples

 Find the derivative of: $y = (2x^3 - 5x^2 + 4)^5$.

 $D_x = \dfrac{d}{dx}$. This problem can be solved by simply applying the theorem for $d(u^n)$. However, to illustrate the use of the chain rule, make the following substitutions:

$$y = u^5 \quad \text{where } u = 2x^3 - 5x^2 + 4$$

Therefore, from the chain rule,

$$D_x y = D_u y \cdot D_x u = 5u^4(6x^2 - 10x)$$

$$= 5(2x^3 - 5x^2 + 4)^4 (6x^2 - 10x).$$

 Find the derivative of: $y = (x^2 + 2)^3$.

 Method 1. We may expand the cube and write:

$$\frac{dy}{dx} = \frac{d}{dx} [(x^2 + 2)^3] = \frac{d}{dx} (x^6 + 6x^4 + 12x^2 + 8)$$

$$= 6x^5 + 24x^3 + 24x.$$

Method 2. Let $u = x^2 + 2$, then $y = (x^2 + 2)^3 = u^3$; Using the chain rule we have:

$$\frac{dy}{dx} = \frac{dy}{du} \cdot \frac{du}{dx} = \frac{d(u^3)}{du} \cdot \frac{d(x^2 + 2)}{dx} = 3u^2(2x)$$

$$= 3(x^2 + 2)^2 \cdot (2x) = 3(x^4 + 4x^2 + 4) \cdot (2x)$$

$$= 6x^5 + 24x^3 + 24x.$$

Find the total differential of the function:
$$z = x^3y + x^2y^2 - 3xy^3.$$

By definition,

$$dz = \frac{\partial z}{\partial x} \, dx + \frac{\partial z}{\partial y} \, dy.$$

$$\frac{\partial z}{\partial x} = 3x^2y + 2xy^2 - 3y^3,$$

and

$$\frac{\partial z}{\partial y} = x^3 + 2x^2y - 9xy^2 \, .$$

Then,

$$dz = (3x^2y + 2xy^2 - 3y^3)dx + (x^3 + 2x^2y - 9xy^2)dy.$$

If $u = x^2 - xy + y^2$, and if $x = 1 + t^2$, $y = 1 - t^2$, what is the total derivative of u?

u is a function of x and y which are each, in turn, functions of t. The total derivative of u with respect to t is given by,

$$\frac{du}{dt} = \frac{\partial u}{\partial x} \frac{dx}{dt} + \frac{\partial u}{\partial y} \frac{dy}{dt} \, .$$

Note that if u were a function of three variables, $u = F(x,y,z)$, which are all, in turn, functions of t, the total derivative of u with respect to t would be written as:

$$\frac{du}{dt} = \frac{\partial u}{\partial x} \frac{dx}{dt} + \frac{\partial u}{\partial y} \frac{dy}{dt} + \frac{\partial u}{\partial z} \frac{dz}{dt} \, .$$

To solve this problem, we need to find $\frac{\partial u}{\partial x}$, $\frac{\partial u}{\partial y}$, $\frac{dx}{dt}$ and $\frac{dy}{dt}$, and then to substitute in the equation for the total derivative, $\frac{du}{dt}$. Hence

$$\frac{\partial u}{\partial y} = 2x - y \qquad\qquad \frac{dx}{dt} = 2t$$

$$\frac{\partial u}{\partial y} = 2y - x \qquad\qquad \frac{dy}{dt} = -2t$$

and

$$\frac{du}{dt} = (2x - y)\,(2t) + (2y - x)\,(-2t) = 2t(3x - 3y) \text{ which, to obtain a}$$

relation in terms of t alone, is:

$$2t[3(1 + t^2) - 3(1 - t^2)]$$

$$= 2t(3 + 3t^2 - 3 + 3t^2)$$

$$= 12t^3.$$

Exercises VII

(See answers on page 299)

You can now successfully try the following.

(1) If $u = \frac{1}{2}x^3$; $v = 3(u + u^2)$; and $w = \dfrac{1}{v^2}$, find $\dfrac{dw}{dx}$

(2) If $y = 3x^2 + \sqrt{2}$; $z = \sqrt{1 + y}$; and $v = \dfrac{1}{\sqrt{3 + 4z}}$, find $\dfrac{dv}{dx}$

(3) If $y = \dfrac{x^3}{\sqrt{3}}$; $z = (1 + y)^2$; and $u = \dfrac{1}{\sqrt{1 + z}}$, find $\dfrac{du}{dx}$

The following exercises are placed here for reasons of space and because their solution depends upon the dodge explained in the foregoing chapter, but they should not be attempted until Chapters 14 and 15 have been read.

(4) If $y = 2a^3\log_e u - u\left(5a^2 - 2au + \frac{1}{3}u^2\right)$, and $u = a + x$,

show that $\qquad\qquad \dfrac{dy}{dx} = \dfrac{x^2(a - x)}{a + x}$

(5) For the curve $x = a(\theta - \sin \theta)$, $y = a(1 - \cos \theta)$, find $\dfrac{dx}{d\theta}$ and $\dfrac{dy}{d\theta}$; hence deduce the value of $\dfrac{dy}{dx}$

(6) Find $\dfrac{dx}{d\theta}$ and $\dfrac{dy}{d\theta}$ for the curve $x = a \cos^3\theta$, $y = a \sin^3\theta$; hence obtain $\dfrac{dy}{dx}$

(7) Given that $y = \log_e\sin (x^2 - a^2)$, find $\dfrac{dy}{dx}$ in its simplest form.

(8) If $u = x + y$ and $4x = 2u - \log_e(2u - 1)$, show that

$$\frac{dy}{dx} = \frac{x+y}{x+y-1}$$

CHAPTER 10

Geometrical Meaning Of Differentiation

It is useful to consider what geometrical meaning can be given to the derivative.

In the first place, any function of x, such, for example, as x^2, or $\sqrt{x}$, or $ax + b$, can be plotted as a curve; and nowadays every student is familiar with the process of curve - plotting.

Let PQR, in Fig. 7, be a portion of a curve plotted with respect to the axes of coordinates OX and OY. Consider any point Q on this curve, where the abscissa of the point is x and its ordinate is y. Now observe how y changes when x is varied. If x is made to increase by a small increment dx, to the right, it will be observed that y also (in *this* particular curve) increases by a small increment dy (because this particular curve happens to be an *ascending* curve). Then the ratio of dy to dx is a measure of the degree to which the curve is sloping up between the two points Q and T. As a matter of fact, it can be seen on the figure that the curve between Q and T has many different slopes, so that we cannot very well speak of the slope of the curve between Q and T. If, however,

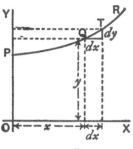

FIG. 7.

Q and T are so near each other that the small portion QT of the curve is practically straight, then it is true to say that the ratio $\dfrac{dy}{dx}$ is the slope of the curve along QT. The straight line QT produced on either side touches the curve along the portion QT only, and if this portion is infinitely small, the straight line will touch the curve at practically one point only, and be therefore a *tangent* to the curve.

This tangent to the curve has evidently the same slope as QT, so that $\dfrac{dy}{dx}$ is the slope of the tangent to the curve at the point Q for which the value of $\dfrac{dy}{dx}$ is found.

We have seen that the short expression "the slope of a curve" has no precise meaning, because a curve has so many slopes—in fact, every small portion of a curve has a different slope. "The slope of a curve *at a point*" is, however, a perfectly defined thing; it is the slope of a very small portion of the curve situated just at that point; and we have seen that this is the same as "the slope of the tangent to the curve at that point".

Observe that dx is a short step to the right, and dy the corresponding short step upwards. These steps must be considered as short as possible—in fact infinitely short,—though in diagrams we have to represent them by bits that are not infinitesimally small, otherwise they could not be seen.

We shall hereafter make considerable use of this circumstance that $\dfrac{dy}{dx}$ *represents the slope of the tangent to the curve at any point.*

If a curve is sloping up at $45°$ at a particular point, as in Fig. 8, dy and dx will be equal, and the value of $\dfrac{dy}{dx} = 1$.

If the curve slopes up steeper than $45°$ (Fig. 9), $\dfrac{dy}{dx}$ will be greater than 1.

83

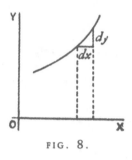

FIG. 8.

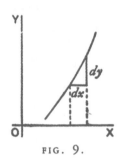

FIG. 9.

If the curve slopes up very gently, as in Fig. 10, $\frac{dy}{dx}$ will be a fraction smaller than 1.

For a horizontal line, or a horizontal place in a curve, $dy = 0$, and therefore

$$\frac{dy}{dx} = 0.$$

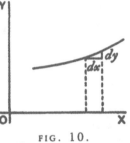

FIG. 10.

If a curve slopes *downward*, as in Fig. 11, dy will be a step down, and must therefore be reckoned of

negative value; hence $\frac{dy}{dx}$ will have a negative sign also.

If the "curve" happens to be a straight line, like that in Fig. 12,

the value of $\frac{dy}{dx}$ will be the same at all points along it. In other

words its *slope* is constant.

If a curve is one that turns more upwards as it goes along to the

right, the values of $\frac{dy}{dx}$ will become greater and greater with the

increasing steepness, as in Fig. 13.

If a curve is one that gets flatter and flatter as it goes along, the

values of $\frac{dy}{dx}$ will become smaller and smaller as the flatter part is

reached, as in Fig. 14.

84

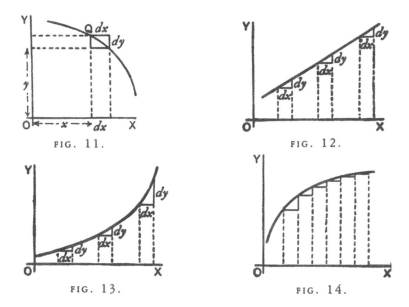

FIG. 11.

FIG. 12.

FIG. 13.

FIG. 14.

If a curve first descends, and then goes up again, as in Fig. 15, presenting a concavity upwards, then clearly $\dfrac{dy}{dx}$ will first be negative, with diminishing values as the curve flattens, then will be zero at the point where the bottom of the trough of the curve is reached; and from this point onward $\dfrac{dy}{dx}$ will have positive values that go on increasing. In such a case y is said to pass through a *minimum*. This value of y is not necessarily the smallest value of y, it is the value of y corresponding to the bottom of the trough; for instance, in Fig. 28, the value of y corresponding to the bottom of the trough is 1, while y takes elsewhere values which are smaller than this. The characteristic of a minimum is that y must increase *on either side* of it.

N.B.—For the particular value of x that makes y *a minimum*, the value of $\dfrac{dy}{dx} = 0$.

85

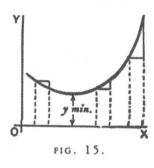

FIG. 15.

If a curve first ascends and then descends, the values of $\frac{dy}{dx}$ will be positive at first; then zero, as the summit is reached; then negative, as the curve slopes downwards, as in Fig. 16.

In this case y is said to pass through a *maximum*, but this value of y is

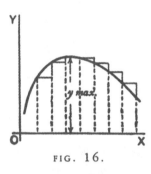

FIG. 16.

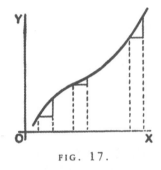

FIG. 17.

not necessarily the greatest value of y. In Fig. 28, the maximum of y is $2\frac{1}{3}$, but this is by no means the greatest value y can have at some other point of the curve.

N.B.—For the particular value of x that makes y *a maximum*, the value of $\frac{dy}{dx} = 0$.

If a curve has the particular form of Fig. 17, the values of $\frac{dy}{dx}$ will always be positive; but there will be one particular place where the slope is least steep, where the value of $\frac{dy}{dx}$ will be a minimum; that is, less than it is at any other part of the curve.

If a curve has the form of Fig. 18, the value of $\frac{dy}{dx}$ will be negative in the upper part, and positive in the lower part; while at the nose of the curve where it becomes actually perpendicular, the value of $\dfrac{dy}{dx}$ will be infinitely great.

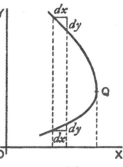

Now that we understand that $\dfrac{dy}{dx}$ measures the steepness of a curve at any point, let us turn to some of the equations which we have already learned how to differentiate.

FIG. 18.

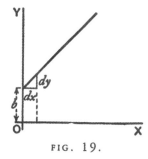

FIG. 19.

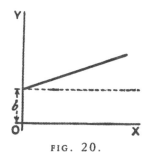

FIG. 20.

(1) As the simplest case take this:

$$y = x + b$$

It is plotted out in Fig. 19, using equal scales for x and y. If we put $x = 0$, then the corresponding ordinate will be $y = b$; that is to say, the "curve" crosses the y-axis at the height b. From here it ascends at 45°; for whatever values we give to x to the right, we have an equal y to ascend. The line has a gradient of 1 in 1.

Now differentiate $y = x + b$, by the rules we have already learned and we get $\dfrac{dy}{dx} = 1$.

The slope of the line is such that for every little step dx to the right, we go an equal little step dy upward. And this slope is constant—always the same slope.

(2) Take another case:

$$y = ax + b.$$

87

We know that this curve, like the preceding one, will start from a height b on the y-axis. But before we draw the curve, let

us find its slope by differentiating; which gives us $\dfrac{dy}{dx} = a$. The

slope will be constant, at an angle, the tangent of which is here called a. Let us assign to a some numerical value—say $\frac{1}{3}$. Then we must give it such a slope that it ascends 1 in 3; or dx will be 3 times as great as dy; as magnified in Fig. 21. So, draw the line in Fig. 20 at this slope.

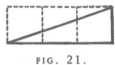

FIG. 21.

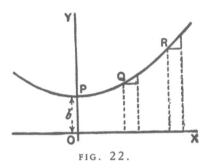

FIG. 22.

(3) Now for a slightly harder case. Let

$$y = ax^2 + b$$

Again the curve will start on the y-axis at a height b above the origin.

Now differentiate. [If you have forgotten, turn back to Chapter 5, or, rather, *don't* turn back, but think out the differentiation.]

$$\frac{dy}{dx} = 2ax$$

This shows that the steepness will not be constant: it increases as x increases. At the starting point P, where $x = 0$, the curve (Fig. 22) has no steepness—that is, it is level. On the left of the origin, where x has negative values, $\dfrac{dy}{dx}$ will also have negative values, or will descend from left to right, as in the Figure.

Let us illustrate this by working out a particular instance. Taking the equation

$$y = \tfrac{1}{4}x^2 + 3$$

and differentiating it, we get

$$\frac{dy}{dx} = \tfrac{1}{2}x$$

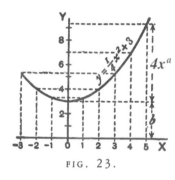

FIG. 23.

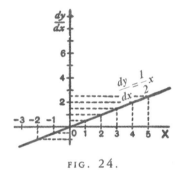

FIG. 24.

Now assign a few successive values, say from 0 to 5, to x; and calculate the corresponding values of y by the first equation; and of $\dfrac{dy}{dx}$ from the second equation. Tabulating results, we have:

x	0	1	2	3	4	5
y	3	$3\tfrac{1}{4}$	4	$5\tfrac{1}{4}$	7	$9\tfrac{1}{4}$
$\dfrac{dy}{dx}$	0	$\tfrac{1}{2}$	1	$1\tfrac{1}{2}$	2	$2\tfrac{1}{2}$

Then plot them out in two curves, Figs. 23 and 24; in Fig. 23 plotting the values of y against those of x, and in Fig. 24 those of

$\dfrac{dy}{dx}$ against those of x. For any assigned value of x, the *height* of the

ordinate in the second curve is proportional to the *slope* of the first curve.

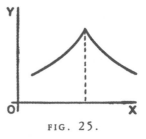

FIG. 25.

If a curve comes to a sudden *cusp*, as in Fig. 25, the slope at that point suddenly changes from a slope upward

to a slope downward. In that case $\dfrac{dy}{dx}$

will clearly undergo an abrupt change from a positive to a negative value.

The following examples show further applications of the principles just explained.

(4) Find the slope of the tangent to the curve $y = \dfrac{1}{2x} + 3$ at the point where $x = -1$. Find the angle which this tangent makes with the curve $y = 2x^2 + 2$.

The slope of the tangent is the slope of the curve at the point where they touch one another; that is, it is the $\dfrac{dy}{dx}$ of the curve for that point. Here $\dfrac{dy}{dx} = -\dfrac{1}{2x^2}$ and for $x = -1$, $\dfrac{dy}{dx} = -\dfrac{1}{2}$, which is the slope of the tangent and of the curve at that point. The tangent, being a straight line, has for equation $y = ax + b$, and its slope is $\dfrac{dy}{dx} = a$, hence $a = -\dfrac{1}{2}$. Also if $x = -1$, $y = \dfrac{1}{2(-1)} + 3 = 2\frac{1}{2}$; and as the tangent passes by this point, the coordinates of the point must satisfy the equation of the tangent, namely

$$y = -\frac{1}{2}x + b$$

so that $2\tfrac{1}{2} = -\dfrac{1}{2} \times (-1) + b$ and $b = 2$; the equation of the tangent

is therefore $y = -\dfrac{1}{2}x + 2$

Now, when two curves meet, the intersection being a point common to both curves, its coordinates must satisfy the equation of each one of the two curves; that is, it must be a solution of the system of simultaneous equations formed by coupling together the equations of the curves. Here the curves meet one another at points given by the solution of

$$\begin{cases} y = 2x^2 + 2 \\ y = -\tfrac{1}{2}x + 2 \end{cases} \quad \text{or} \quad 2x^2 + 2 = -\tfrac{1}{2}x + 2$$

that is, $$x\!\left(2x + \tfrac{1}{2}\right) = 0$$

This equation has for its solutions $x = 0$ and $x = -\tfrac{1}{4}$. The slope of the curve $y = 2x^2 + 2$ at any point is

$$\frac{dy}{dx} = 4x$$

For the point where $x = 0$, this slope is zero; the curve is horizontal. For the point where

$$x = -\frac{1}{4}, \quad \frac{dy}{dx} = -1$$

hence the curve at that point slopes downwards to the right at such an angle θ with the horizontal that $\tan \theta = 1$; that is, at $45°$ to the horizontal.

The slope of the straight line is $-\tfrac{1}{2}$; that is, it slopes downwards to the right and makes with the horizontal an angle ϕ such that $\tan \phi = \tfrac{1}{2}$; that is, an angle of $26° \ 34'$. It follows that at the first point the curve cuts the straight line at an angle of $26° \ 34'$, while at the second it cuts it at an angle of $45° - 26° \ 34' = 18° \ 26'$.

(5) A straight line is to be drawn, through a point whose co-ordinates are $x = 2$, $y = -1$, as tangent to the curve

$$y = x^2 - 5x + 6$$

Find the coordinates of the point of contact.

The slope of the tangent must be the same as the $\dfrac{dy}{dx}$ of the curve; that is, $2x - 5$.

The equation of the straight line is $y = ax + b$, and as it is satisfied for the values $x = 2$, $y = -1$, then $-1 = a \times 2 + b$; also, its

$$\frac{dy}{dx} = a = 2x - 5$$

The x and the y of the point of contact must also satisfy both the equation of the tangent and the equation of the curve.

We have then

$$\left\{ \begin{array}{ll} y = x^2 - 5x + 6, & \text{.............................(i)} \\ y = ax + b, & \text{...(ii)} \\ -1 = 2a + b, & \text{.......................................(iii)} \\ a = 2x - 5, & \text{...(iv)} \end{array} \right.$$

four equations in a, b, x, y.

Equations (i) and (ii) give $x^2 - 5x + 6 = ax + b$.

Replacing a and b by their value in this, we get

$$x^2 - 5x + 6 = (2x - 5)x - 1 - 2(2x - 5)$$

which simplifies to $x^2 - 4x + 3 = 0$, the solutions of which are: $x = 3$ and $x = 1$. Replacing in (i), we get $y = 0$ and $y = 2$ respectively; the two points of contact are then $x = 1$, $y = 2$; and $x = 3$, $y = 0$.

Note.—In all exercises dealing with curves, students will find it extremely instructive to verify the deductions obtained by actually plotting the curves.

Additional Problem Solving Examples

Find the maxima and minima of $f(x) = 3x^5 - 5x^3$.

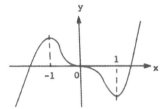

To determine maxima and minima we find $f'(x)$, set it equal to 0 and solve for x, obtaining the critical points. We find:

$$f(x) = 15x^4 - 15x^2 = 15x^2(x^2 - 1).$$

Therefore, $x = 0, \pm 1$ are the critical points. We must now determine whether the function reaches a maximum, minimum or neither at each of these values. To do this we will use the Second Derivative Test. Computing the second derivative f'', we have:

$$f''(x) = 60x^3 - 30x = 30(2x^2 - 1).$$

$f''(1) = 30 > 0$. Therefore, $x = 1$ is a relative minimum point and $x = -1$ is a relative maximum since $f''(-1) = -30 < 0$. Now, $f''(0) = 0$. Therefore the Second Derivative Test indicates a point which is neither maximum nor minimum. This is known as a point of inflection. For further study of the behavior of f at 0 we must use the First Derivative Test. We examine $f'(x)$ when $-1 < x < 0$ and when $0 < x < 1$. Let us select a representative value from each interval. We will use

$x = -\dfrac{1}{2}$ and $x = \dfrac{1}{2}$. For $f'\left(-\dfrac{1}{2}\right)$ we obtain a negative value, and

for $f'\left(\dfrac{1}{2}\right)$ we again obtain a negative value. Because there is no

change in sign we conclude that at $x = 0$ there is neither a maximum nor a minimum, as can also be seen from the graph.

93

Concavity

If a function is differentiable on an open interval containing c, then the graph at this point is

a) concave upward (or convex) if f"(c) > 0;

b) concave downward if f"(c) < 0.

If a function is concave upward than f' is increasing as x increases. If the function is concave downward, f' is decreasing as x increases.

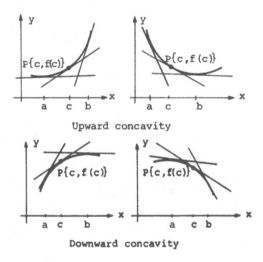

Upward concavity

Downward concavity

 Find the intervals of x for which the curve

$$y = 2x^3 - 9x^2 + 12x - 3$$

is concave downward and concave upward.

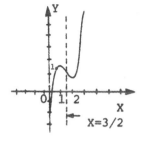

Differentiating twice,

$$\frac{dy}{dx} = 6(x^2 - 3x + 2)$$

and

$$\frac{d^2y}{dx^2} = 6(2x - 3)$$

By setting $\frac{d^2y}{dx^2} = 0$, we have $x = \frac{3}{2}$

y'' is positive or negative according to $x > \frac{3}{2}$ Or $< \frac{3}{2}$. Hence the graph is concave downward to the left of $x = \frac{3}{2}$ and concave upward to the right of $x = \frac{3}{2}$.

Points of Inflection

Points which satisfy $f''(x) = 0$ may be positions where concavity changes. These points are called the points of inflection. It is the point at which the curve crosses its tangent line.

Investigate the function $y = x - a^{\frac{1}{3}} \; 2x - a^{\frac{2}{3}}$ for maxima and minima.

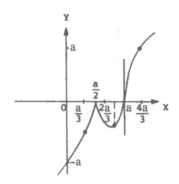

Differentiating

$$y' = \frac{(2x-a)^{\frac{2}{3}}}{3(x-a)^{\frac{2}{3}}} + \frac{4(x-a)^{\frac{1}{3}}}{3(2x-a)^{\frac{1}{3}}} = \frac{6x - 5a}{3(x-a)^{\frac{2}{3}} \; (2x-a)^{\frac{1}{3}}} \; .$$

From $y' = 0$, and $\frac{1}{y'} = 0$, the critical points are $x = \frac{a}{2}, x = \frac{5a}{6}, x = a$.

We must now determine whether each of these critical points is a maximum, minimum, or neither. We choose a value of x less than and a value greater than each of the critical values and evaluate y' at these values. If the sign changes from positive to negative, we have a maximum. If it changes from negative to positive, we have a minimum. If the sign does not change, there is neither one at that critical value.

Setting $x = \frac{a}{3}$ and $\frac{2a}{3}$ in turn, we have y' positive and negative respectively. Hence $x = \frac{a}{2}$ makes y a maximum.

Test $x = \frac{5a}{6}$, using the values $\frac{2a}{3}$ and $\frac{9a}{10}$.

These show y′ to be successively negative and positive, so the function has a minimum value at x = $\dfrac{5a}{6}$.

Apply the test to x = a, with the values $\dfrac{9a}{10}$ and 2a. These show y′ positive in both cases, therefore, at x = a there is neither a maxi-

mum or a minimum. We observe that y′ = $\dfrac{a}{0}$ = ∞ at x = a,

therefore the graph has a vertical tangent at this value. At x = a there

is a point of inflection as shown. The maximum point at ($\dfrac{a}{2}$, 0) is called a cusp.

Exercises VIII

(See answers on page 300)

(1) Plot the curve $y = \frac{3}{4}x^2 - 5$, using a scale of millimeters. Measure at points corresponding to different values of x, the angle of its slope.

Find, by differentiation the equation, the expression for slope; and see from a table of tangents in your calculator whether this agrees with the measured angle.

(2) Find what will be the slope of the curve

$$y = 0.12x^3 - 2$$

at the point $x = 2$

(3) If $y = (x - a)(x - b)$, show that at the particular point of the curve where $\dfrac{dy}{dx} = 0$, x will have the value $\frac{1}{2}(a + b)$.

(4) Find the $\dfrac{dy}{dx}$ of the equation $y = x^3 + 3x$; and calculate the numerical values of $\dfrac{dy}{dx}$ for the points corresponding to $x = 0$, $x = \frac{1}{2}$, $x = 1$, $x = 2$.

(5) In the curve to which the equation is $x^2 + y^2 = 4$, find the

value of x at those points where the slope $= 1$.

(6) Find the slope, at any point, of the curve whose equation is $\dfrac{x^2}{3^2} + \dfrac{y^2}{2^2} = 1$; and give the numerical value of the slope at the place where $x = 0$, and at that where $x = 1$.

(7) The equation of a tangent to the curve $y = 5 - 2x + 0.5x^3$, being of the form $y = mx + n$, where m and n are constants, find the value of m and n if the point where the tangent touches the curve has $x = 2$ for abscissa.

(8) At what angle do the two curves

$$y = 3.5x^2 + 2 \quad \text{and} \quad y = x^2 - 5x + 9.5$$

cut one another?

(9) Tangents to the curve $y = \pm\sqrt{25 - x^2}$ are drawn at points for which $x = 3$ and $x = 4$, the value of y being positive. Find the coordinates of the point of intersection of the tangents and their mutual inclination.

(10) A straight line $y = 2x - b$ touches a curve $y = 3x^2 + 2$ at one point. What are the coordinates of the point of contact, and what is the value of b?

CHAPTER 11

Maxima And Minima

Why We Study Maxima and Minima, and How this Topic Is Related to Differentiation

An important goal in business is to obtain maximum profits. In science and engineering it is often necessary, for example, to calculate maximum and minimum speeds, accelerations, production quantities, efficiencies, power requirements, and working or operating times.

When any of the preceding variables is plotted or its function is known, then maximum and minimum points can be found by applying differentiation.

A quantity which varies continuously is said to pass by (or through) a local maximum or minimum value when, in the course of its variation, the immediately preceding and following values are *both* smaller or greater, respectively, than the value referred to. An infinitely great value is therefore not a maximum value.

One of the principal uses of the process of differentiating is to find out under what conditions the value of the thing differentiated becomes a maximum or a minimum. This is often exceedingly important in engineering and financial questions, where it is most desirable to know what conditions will make the cost of working a minimum, or will make the efficiency a maximum.

Now, to begin with a concrete case, let us take the equation

$$y = x^2 - 4x + 7$$

By assigning a number of successive values to x, and finding the corresponding values of y, we can readily see that the equation represents a curve with a minimum.

x	0	1	2	3	4	5
y	7	4	3	4	7	12

These values are plotted in Fig. 26, which shows that y has apparently a minimum value of 3, when x is made equal to 2. But are you sure that the minimum occurs at 2, and not at $2\frac{1}{4}$ or at $1\frac{3}{4}$?

Of course it would be possible with any algebraic expression to work out a lot of values, and in this way arrive gradually at the particular value that may be a maximum or a minimum.

Here is another example: Let

$$y = 3x - x^2$$

Calculate a few values thus:

x	−1	0	1	2	3	4	5
y	−4	0	2	2	0	−4	−10

Plot these values as in Fig. 27.

It will be evident that there will be a maximum somewhere between $x = 1$ and $x = 2$; and the thing *looks* as if the maximum value of y ought to be about $2\frac{1}{4}$. Try some intermediate values. If $x = 1\frac{1}{4}$, $y = 2.187$; if $x = 1\frac{1}{2}$, $y = 2.25$; if $x = 1.6$, $y = 2.24$. How can we be sure that 2.25 is the real maximum, or that it occurs exactly when $x = 1\frac{1}{2}$?

Now it may sound like juggling to be assured that there is a way by which one can arrive straight at a maximum (or minimum) value without making a lot of preliminary trials or guesses. And that way depends on differentiating. Look back to Chapter 10 for the remarks about Figs. 14 and 15, and you will see that whenever a curve gets either to its maximum or to its

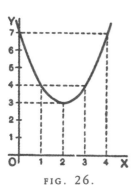

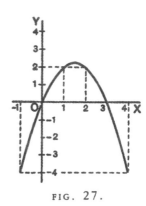

FIG. 26.   FIG. 27.

minimum height, at that point its $\dfrac{dy}{dx} = 0$. Now this gives us the

clue to the dodge that is wanted. When there is put before you an equation, and you want to find that value of x that will make its y a minimum (or a maximum), *first differentiate it,* and having

done so, write its $\dfrac{dy}{dx}$ as *equal to zero,* and then solve for x. Put this

particular value of x into the original equation, and you will then get the required value of y. This process is commonly called "equating to zero".

To see how simply it works, take the example with which this chapter opens, namely,

$$y = x^2 - 4x + 7$$

Differentiating, we get:

$$\frac{dy}{dx} = 2x - 4$$

Now equate this to zero, thus:

101

$$2x - 4 = 0$$

Solving this equation for x, we get:

$$2x = 4, \quad x = 2$$

Now, we know that the maximum (or minimum) will occur exactly when $x = 2$.

Putting the value $x = 2$ into the original equation, we get

$$y = 2^2 - (4 \times 2) + 7$$
$$= 4 - 8 + 7 = 3$$

Now look back at Fig. 26, and you will see that the minimum occurs when $x = 2$, and that this minimum of $y = 3$.

Try the second example (Fig. 27), which is

$$y = 3x - x^2$$

Differentiating, $\qquad \dfrac{dy}{dx} = 3 - 2x$

Equating to zero, $\quad 3 - 2x = 0$

whence $\qquad\qquad\qquad x = 1\frac{1}{2}$

and putting this value of x into the original equation, we find:

$$y = 4\tfrac{1}{2} - \left(1\tfrac{1}{2} \times 1\tfrac{1}{2}\right)$$
$$y = 2\tfrac{1}{4}$$

This gives us exactly the information as to which the method of trying a lot of values left us uncertain.

Now, before we go on to any further cases, we have two remarks to make. When you are told to equate $\dfrac{dy}{dx}$ to zero, you feel at first (that is if you have any wits of your own) a kind of resentment, because you know that $\dfrac{dy}{dx}$ has all sorts of different values at different parts of the curve, according to whether it is sloping up or down. So, when you are suddenly told to write

$$\frac{dy}{dx} = 0$$

you resent it, and feel inclined to say that it can't be true. Now you will have to understand the essential difference between "an equation", and "an equation of condition". Ordinarily you are dealing with equations that are true in themselves; but, on occasions, of which the present are examples, you have to write down equations that are not necessarily true, but are only true if certain conditions are to be fulfilled; and you write them down in order, by solving them, to find the conditions which make them true. Now we want to find the particular value that x has when the curve is neither sloping up nor sloping down, that is, at the particular place where $\dfrac{dy}{dx} = 0$. So, writing $\dfrac{dy}{dx} = 0$ does *not* mean that it always is $= 0$; but you write it down *as a condition* in order to see how much x will come out if $\dfrac{dy}{dx}$ is to be zero.

The second remark is one which (if you have any wits of your own) you will probably have already made: namely, that this much-belauded process of equating to zero entirely fails to tell you whether the x that you thereby find is going to give you a *maximum* value of y or a *minimum* value of y. Quite so. It does not of itself discriminate; it finds for you the right value of x but leaves you to find out for yourselves whether the corresponding y is a maximum or a minimum. Of course, if you have plotted the curve, you know already which it will be.

For instance, take the equation:

$$y = 4x + \frac{1}{x}$$

Without stopping to think what curve it corresponds to, differentiate it, and equate to zero:

$$\frac{dy}{dx} = 4 - x^{-2} = 4 - \frac{1}{x^2} = 0$$

whence
$$x = \pm\frac{1}{2}$$

and, inserting these values,
$$y = \pm 4$$

Each will be either a maximum or else a minimum. But which? You will hereafter be told a way, depending upon a second differentiation (see Chapter 12). But at present it is enough if you will simply try two other values of x differing a little from the one found, one larger and one smaller, and see whether with these altered values the corresponding values of y are less or greater than that already found.

Try another simple problem in maxima and minima. Suppose you were asked to divide any number into two parts, such that the product was a maximum? How would you set about it if you did not know the trick of equating to zero? I suppose you could worry it out by the rule of try, try, try again. Let 60 be the number. You can try cutting it into two parts, and multiplying them together. Thus, 50 times 10 is 500; 52 times 8 is 416; 40 times 20 is 800; 45 times 15 is 675; 30 times 30 is 900. This looks like a maximum: try varying it. 31 times 29 is 899, which is not so good; and 32 times 28 is 896, which is worse. So it seems that the biggest product will be got by dividing into two halves.

Now see what the calculus tells you. Let the number to be cut into two parts be called n. Then if x is one part, the other will be $n - x$, and the product will be $x(n - x)$ or $nx - x^2$. So we write $y = nx - x^2$. Now differentiate and equate to zero;

$$\frac{dy}{dx} = n - 2x = 0$$

Solving for x, we get
$$\frac{n}{2} = x$$

104

So now we *know* that whatever number n may be, we must divide it into two equal parts if the product of the parts is to be a maximum; and the value of that maximum product will always be $= \frac{1}{4}n^2$.

This is a very useful rule, and applies to any number of factors, so that if $m + n + p =$ a constant number, $m \times n \times p$ is a maximum when $m = n = p$.

Test Case.

Let us at once apply our knowledge to a case that we can test.

Let $$y = x^2 - x$$

and let us find whether this function has a maximum or minimum; and if so, test whether it is a maximum or a minimum.

Differentiating, we get

$$\frac{dy}{dx} = 2x - 1$$

Equating to zero, we get

$$2x - 1 = 0$$

whence $$2x = 1$$

or $$x = \frac{1}{2}$$

That is to say, when x is made $= \frac{1}{2}$, the corresponding value of y will be either a maximum or a minimum. Accordingly, putting $x = \frac{1}{2}$ in the original equation, we get

$$y = \left(\tfrac{1}{2}\right)^2 - \tfrac{1}{2}$$

or $$y = -\tfrac{1}{4}$$

Is this a maximum or a minimum? To test it, try putting x a little bigger than $\frac{1}{2}$—say, make $x = 0.6$. Then

$$y = (0.6)^2 - 0.6 = 0.36 - 0.6 = -0.24$$

which is higher than -0.25, showing that $y = -0.25$ is a *minimum*.

Plot the curve for yourself, and verify the calculation.

Further Examples.

A most interesting example is afforded by a curve that has both a maximum and a minimum. Its equation is:

$$y = \tfrac{1}{3}x^3 - 2x^2 + 3x + 1$$

Now
$$\frac{dy}{dx} = x^2 - 4x + 3$$

Equating to zero, we get the quadratic,

$$x^2 - 4x + 3 = 0$$

and solving the quadratic gives us *two* roots, viz.

$$\begin{cases} x = 3 \\ x = 1 \end{cases}$$

Now, when $x = 3$, $y = 1$; and when $x = 1$, $y = 2\tfrac{1}{3}$. The first of these is a minimum, the second a maximum.

The curve itself may be plotted (as in Fig. 28) from the values calculated, as below, from the original equation.

x	-1	0	1	2	3	4	5	6
y	$-4\tfrac{1}{3}$	1	$2\tfrac{1}{3}$	$1\tfrac{2}{3}$	1	$2\tfrac{1}{3}$	$7\tfrac{2}{3}$	19

A further exercise in maxima and minima is afforded by the following example:

The equation of a circle of radius r, having its center C at the point whose coordinates are $x = a$, $y = b$, as depicted in Fig. 29, is:

$$(y - b)^2 + (x - a)^2 = r^2$$

This may be transformed into

$$y = \sqrt{r^2 - (x - a)^2} + b$$

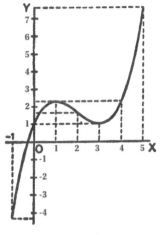

FIG. 28.

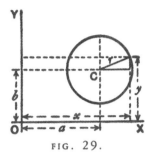

FIG. 29.

Now we know beforehand, by mere inspection of the figure, that when $x = a$, y will be either at its maximum value, $b + r$, or else at its minimum value, $b - r$. But let us not take advantage of this knowledge; let us set about finding what value of x will make y a maximum or a minimum, by the process of differentiating and equating to zero.

$$\frac{dy}{dx} = \frac{1}{2} \frac{1}{\sqrt{r^2 - (x - a)^2}} \times (2a - 2x)$$

which reduces to

$$\frac{dy}{dx} = \frac{a - x}{\sqrt{r^2 - (x - a)^2}}$$

Then the condition for y being maximum or minimum is:

$$\frac{a - x}{\sqrt{r^2 - (x - a)^2}} = 0$$

107

Since no value whatever of x will make the denominator infinite, the only condition to give zero is

$$x = a$$

Inserting this value in the original equation for the circle, we find

$$y = \sqrt{r^2} + b$$

and as the root of r^2 is either $+r$ or $-r$, we have two resulting values of y,

$$\begin{cases} y = b + r \\ y = b - r \end{cases}$$

The first of these is the maximum, at the top; the second the minimum, at the bottom.

If the curve is such that there is no place that is a maximum or minimum, the process of equating to zero will yield an impossible result. For instance:

Let $$y = ax^3 + bx + c$$

Then $$\frac{dy}{dx} = 3ax^2 + b$$

Equating this to zero, we get $3ax^2 + b = 0$, $x^2 = \dfrac{-b}{3a}$, and $x = \sqrt{\dfrac{-b}{3a}}$, which is impossible, supposing a and b to have the same sign.

Therefore y has no maximum nor minimum.

A few more worked examples will enable you to thoroughly master this most interesting and useful application of the calculus.

(1) What are the sides of the rectangle of maximum area inscribed in a circle of radius R?

If one side be called x,

$$\text{the other side} = \sqrt{(\text{diagonal})^2 - x^2};$$

and as the diagonal of the rectangle is necessarily a diameter of the circumscribing circle, the other side $=\sqrt{4R^2 - x^2}$.

Then, area of rectangle $S = x\sqrt{4R^2 - x^2}$,

$$\frac{dS}{dx} = x \times \frac{d\left(\sqrt{4R^2 - x^2}\right)}{dx} + \sqrt{4R^2 - x^2} \times \frac{d(x)}{dx}$$

If you have forgotten how to differentiate $\sqrt{4R^2 - x^2}$, here is a hint: write $w = 4R^2 - x^2$ and $y = \sqrt{w}$, and seek $\dfrac{dy}{dw}$ and $\dfrac{dw}{dx}$;

fight it out, and only if you can't get on refer to Chapter 9.

You will get

$$\frac{dS}{dx} = x \times -\frac{x}{\sqrt{4R^2 - x^2}} + \sqrt{4R^2 - x^2} = \frac{4R^2 - 2x^2}{\sqrt{4R^2 - x^2}}$$

For maximum or minimum we must have

$$\frac{4R^2 - 2x^2}{\sqrt{4R^2 - x^2}} = 0$$

that is, $4R^2 - 2x^2 = 0$ and $x = R\sqrt{2}$.

The other side $= \sqrt{4R^2 - 2R^2} = R\sqrt{2}$; the two sides are equal; the figure is a square the side of which is equal to the diagonal of the square constructed on the radius. In this case it is, of course, a maximum with which we are dealing.

(2) What is the radius of the opening of a conical vessel the sloping side of which has a length l when the capacity of the vessel is greatest?

If R be the radius and H the corresponding height,

$$H = \sqrt{l^2 - R^2}$$

$$\text{Volume } V = \pi R^2 \times \frac{H}{3} = \pi R^2 \times \frac{\sqrt{l^2 - R^2}}{3}$$

109

Proceeding as in the previous problem, we get

$$\frac{dV}{dR} = \pi R^2 \times -\frac{R}{3\sqrt{l^2 - R^2}} + \frac{2\pi R}{3}\sqrt{l^2 - R^2}$$

$$= \frac{2\pi R(l^2 - R^2) - \pi R^3}{3\sqrt{l^2 - R^2}} = 0$$

for maximum or minimum.

Or, $2\pi R(l^2 - R^2) - \pi R^3 = 0$, and $R = l\sqrt{\frac{2}{3}}$, for a maximum, obviously.

(3) Find the maxima and minima of the function

$$y = \frac{x}{4 - x} + \frac{4 - x}{x}$$

We get $\quad \dfrac{dy}{dx} = \dfrac{(4 - x) - (-x)}{(4 - x)^2} + \dfrac{-x - (4 - x)}{x^2} = 0$

for maximum or minimum; or

$$\frac{4}{(4 - x)^2} - \frac{4}{x^2} = 0 \quad \text{and} \quad x = 2$$

There is only one value, hence only one maximum or minimum.

$$\text{For} \quad x = 2 \quad y = 2$$
$$\text{for} \quad x = 1.5 \quad y = 2.27$$
$$\text{for} \quad x = 2.5 \quad y = 2.27$$

it is therefore a minimum. (It is instructive to plot the graph of the function.)

(4) Find the maxima and minima of the function

$$y = \sqrt{1 + x} + \sqrt{1 - x}$$

(It will be found instructive to plot the graph.)

Differentiating gives at once (see example No. 1, Chapter 9).

$$\frac{dy}{dx} = \frac{1}{2\sqrt{1 + x}} - \frac{1}{2\sqrt{1 - x}} = 0$$

for maximum or minimum.

Hence $\sqrt{1+x} = \sqrt{1-x}$ and $x = 0$, the only solution

For $x = 0$, $y = 2$.

For $x = \pm 0.5$, $y = 1.932$, so this is a maximum.

(5) Find the maxima and minima of the function

$$y = \frac{x^2 - 5}{2x - 4}$$

We have $\dfrac{dy}{dx} = \dfrac{(2x-4) \times 2x - (x^2 - 5)2}{(2x-4)^2} = 0$

for maximum or minimum; or

$$\frac{2x^2 - 8x + 10}{(2x - 4)^2} = 0$$

or $x^2 - 4x + 5 = 0$; which has solutions

$$x = 2 \pm \sqrt{-1}$$

These being imaginary, there is no real value of x for which $\dfrac{dy}{dx} = 0$; hence there is neither maximum nor minimum.

(6) Find the maxima and minima of the function

$$(y - x^2)^2 = x^5$$

This may be written $y = x^2 \pm x^{\frac{5}{2}}$.

$$\frac{dy}{dx} = 2x \pm \tfrac{5}{2}x^{\frac{3}{2}} = 0 \text{ for maximum or minimum;}$$

that is, $x\left(2 \pm \tfrac{5}{2}x^{\frac{1}{2}}\right) = 0$, which is satisfied for $x = 0$, and for $2 \pm \tfrac{5}{2}x^{\frac{1}{2}} = 0$, that is for $x = \tfrac{16}{25}$. So there are two solutions.

Taking first $x = 0$. If $x = -0.5$, $y = 0.25 \pm \sqrt[2]{-(.5)^5}$, and if $x = +0.5$, $y = 0.25 \pm \sqrt[2]{(.5)^5}$. On one side y is imaginary; that is, there is no value of y that can be represented by a graph; the

111

latter is therefore entirely on the right side of the axis of y (see Fig. 30).

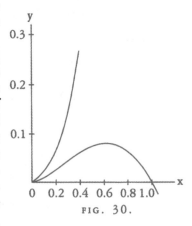

FIG. 30.

On plotting the graph it will be found that the curve goes to the origin, as if there were a minimum there; but instead of continuing beyond, as it should do for a minimum, it retraces its steps (forming a cusp). There is no minimum, therefore, although the condition for a minimum is satisfied, namely $\frac{dy}{dx} = 0$. It is necessary therefore always to check by taking one value on either side.

Now, if we take $x = \frac{16}{25} = 0.64$. If $x = 0.64$, $y = 0.7373$ and $y = 0.0819$; if $x = 0.6$, y becomes 0.6389 and 0.0811; and if $x = 0.7$, y becomes 0.9000 and 0.0800.

This shows that there are two branches of the curve; the upper one does not pass through a maximum, but the lower one does.

(7) A cylinder whose height is twice the radius of the base is increasing in volume, so that all its parts keep always in the same proportion to each other; that is, at any instant, the cylinder is *similar* to the original cylinder. When the radius of the base is r inches, the surface area is increasing at the rate of 20 square inches per second; at what rate per second is its volume then increasing?

$$\text{Area} = S = 2(\pi r^2) + 2\pi r \times 2r = 6\pi r^2$$

$$\text{Volume} = V = \pi r^2 \times 2r = 2\pi r^3$$

$$\frac{dS}{dt} = 12\pi r\frac{dr}{dt} = 20; \quad \frac{dr}{dt} = \frac{20}{12\pi r}$$

$$\frac{dV}{dt} = 6\pi r^2\frac{dr}{dt}; \quad \text{and}$$

$$\frac{dV}{dt} = 6\pi r^2 \times \frac{20}{12\pi r} = 10r$$

The volume changes at the rate of 10r cubic inches per second.

Make other examples for yourself. There are few subjects which offer such a wealth for interesting examples.

Additional Problem Solving Examples

 Find the maxima and minima of the function $f(x) = x^4$.

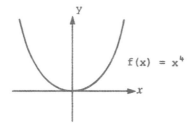

To determine maxima and minima we find $f'(x)$, set it equal to 0, and solve for x to obtain the critical points. We find: $f'(x) = 4x^3 = 0$, therefore x = 0 is the critical value. We must now determine whether x = 0 is a maximum or minimum value. In this example the Second Derivative Test fails because $f''(x) = 12x^2$ and $f''(0) = 0$. We must, therefore, use the First Derivative Test. We examine $f'(x)$ when x < 0, and when x > 0. We find that for x < 0, $f'(x)$ is negative, and for x > 0, $f'(x)$ is positive. Therefore there is a minimum at (0, 0). (See figure).

 Locate the maxima and minima of $y = 2x^2 - 8x + 6$.

 To obtain the minima and maxima we find $\dfrac{dy}{dx}$, set it equal to 0 and solve for x. We find:

$$\frac{dy}{dx} = 4x - 8 = 0.$$

Therefore, $x = 2$ is the critical point. We now use the Second Derivative Test to determine whether $x = 2$ is a maximum or a minimum. We find :

$$\frac{d^2y}{dx^2} = 4, \text{(positive). The second derivative is positive, hence } x = 2 \text{ is}$$

a minimum.

Now substitute this back into the original equation to get the corresponding ordinate.

$$y = 2x^2 - 8x + 6 = 2 \cdot 2^2 - 8 \cdot 2 + 6 = 8 - 16 + 6$$

$$= -2.$$

Therefore, the minimum is at $x = 2$, $y = -2$.

 Locate the maxima and minima of

$$y = \frac{x^3}{3} - \frac{5x^2}{2} + 6x + 4.$$

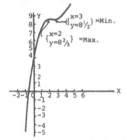

 To find the maxima and minima we find $\dfrac{dy}{dx}$, set it to 0, and solve for x, obtaining the critical points. Doing this we have:

$$\frac{dy}{dx} = x^2 - 5x + 6 = 0, \quad (x - 2)(x - 3) = 0$$

therefore,

$$x = 3 \text{ and } 2.$$

We now use the Second Derivative Test to determine whether the critical values are maximum, minimum or neither. We find:

$$\frac{d^2y}{dx^2} = 2x - 5.$$

For x = 3,

$$\frac{d^2y}{dx^2} = 2x - 5 = 2 \cdot 3 - 5 = + \text{(positive)},$$

which indicates a minimum.

For x = 2,

$$\frac{d^2y}{dx^2} = 2x - 5 = 2 \cdot 2 - 5 = - \text{(negative)},$$

which indicates a maximum.

Therefore, we have a minimum at x = 3 and a maximum at x = 2. We now wish to find the corresponding ordinates. Going back to the original equation, we have:

For x = 3,

$$y = \frac{x^3}{3} - \frac{5x^2}{2} + 6x + 4 = \frac{3^3}{3} - \frac{5 \cdot 3^2}{2} + 6 \cdot 3 + 4$$

$$= 9 - \frac{45}{2} + 18 + 4 = 8\frac{1}{2}.$$

For x = 2,

$$y = \frac{x^3}{3} - \frac{5x^2}{2} + 6x + 4 = \frac{2^3}{3} - \frac{5 \cdot 2^2}{2} + 6 \cdot 2 + 4$$

$$= \frac{8}{3} - 10 + 12 + 4 = 8\frac{2}{3}.$$

Therefore, minimum is at x = 3, $y = 8\frac{1}{2}$, and maximum is at x = 2, $y = 8\frac{2}{3}$.

Exercises IX

(See answers on page 300)

(1) What values of x will make y a maximum and a minimum, if $y = \dfrac{x^2}{x+1}$?

(2) What value of x will make y a maximum in the equation $y = \dfrac{x}{a^2 + x^2}$?

(3) A line of length p is to be cut up into 4 parts and put together as a rectangle. Show that the area of the rectangle will be a maximum if each of its sides is equal to $\frac{1}{4}p$.

(4) A piece of string 30 inches long has its two ends joined together and is stretched by 3 pegs so as to form a triangle. What is the largest triangular area that can be enclosed by the string?

(5) Plot the curve corresponding to the equation

$$y = \frac{10}{x} + \frac{10}{8-x}$$

also find $\dfrac{dy}{dx}$, and deduce the value of x that will make y a minimum; and find that minimum value of y.

(6) If $y = x^5 - 5x$, find what values of x will make y a maximum or a minimum.

(7) What is the smallest square that can be inscribed in a given square?

(8) Inscribe in a given cone, the height of which is equal to the radius of the base, a cylinder (*a*) whose volume is a maximum; (*b*) whose lateral area is a maximum; (*c*) whose total area is a maximum.

(9) Inscribe in a sphere, a cylinder (*a*) whose volume is a maximum; (*b*) whose lateral area is a maximum; (*c*) whose total area is a maximum.

(10) A spherical balloon is increasing in volume. If, when its radius is r feet, its volume is increasing at the rate of 4 cubic feet per second, at what rate is its surface then increasing?

(11) Inscribe in a given sphere a cone whose volume is a maximum.

116

CHAPTER 12

Curvature of Curves

We have seen how to apply the second derivative to find accelerations. This chapter describes additional applications of the second derivative to find characteristics of curves and functions, and to specify further maximums and minimums obtained from first derivatives.

Returning to the process of successive differentiation, it may be asked: Why does anybody want to differentiate twice over? We know that when the variable quantities are space and time, by differentiating twice over we get the acceleration of a moving body, and that in the geometrical interpretation, as applied to curves, $\dfrac{dy}{dx}$ means the *slope* of the curve. But what can $\dfrac{d^2y}{dx^2}$ mean in this case? Clearly it means the rate (per unit of length x) at which the slope is changing—in brief, it is *an indication of the manner in which the slope of the portion of curve considered varies,* that is, whether the slope of the curve increases or decreases when x increases, or, in other words, whether the curve curves up or down towards the right.

Suppose a slope constant, as in Fig. 31.

Here, $\dfrac{dy}{dx}$ is of constant value.

Suppose, however, a case in which, like Fig. 32, the slope itself

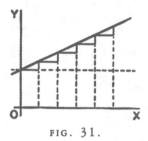

FIG. 31.

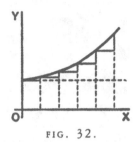

FIG. 32.

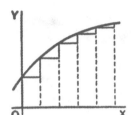

FIG. 33.

is getting greater upwards; then $\dfrac{d\left(\dfrac{dy}{dx}\right)}{dx}$,

that is, $\dfrac{d^2y}{dx^2}$, will be *positive*.

If the slope is becoming less as you go to the right (as in Fig. 14), or as in Fig. 33, then, even though the curve may be going upward, since the change is such as to diminish its slope, its $\dfrac{d^2y}{dx^2}$ will be *negative*.

It is now time to initiate you into another secret—how to tell whether the result that you get by "equating to zero" is a maximum or a minimum. The trick is this: After you have differentiated (so as to get the expression which you equate to zero), you then differentiate a second time and look whether the result of the second differentiation is *positive* or *negative*. If $\dfrac{d^2y}{dx^2}$ comes out *positive*, then you know that the value of y which you got was a *minimum*; but if $\dfrac{d^2y}{dx^2}$ comes out *negative*, then the value of y which you got must be a *maximum*. That's the rule.

The reason of it ought to be quite evident. Think of any curve that has a minimum point in it, like Fig. 15, or like Fig. 34, where the point of minimum y is marked M, and the curve is *concave* upward. To the left of M the slope is downward, that is, negative, and is getting less negative.

To the right of *M* the slope has become upward, and is getting more and more upward. Clearly the change of slope as the curve passes through *M* is such that

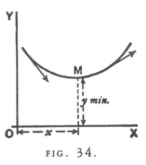

FIG. 34.

$\dfrac{d^2y}{dx^2}$ is *positive*, for its operation, as *x* in-

creases toward the right, is to convert a downward slope into an upward one.

Similarly, consider any curve that has a maximum point in it, like Fig. 16, or like Fig. 35 where the curve is convex when viewed from above and the maximum point is marked M. In this case, as the curve passes through M from left to right, its upward slope is converted into a downward or negative slope, so that in this case the "slope of the slope"

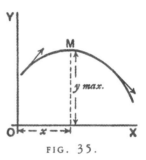

FIG. 35.

$\dfrac{d^2y}{dx^2}$ is *negative*.

Go back now to the examples of the last chapter and verify in this way the conclusions arrived at as to whether in any particular case there is a maximum or a minimum. You will find below a few worked-out examples.

(1) Find the maximum or minimum of

$$(a)\ y = 4x^2 - 9x - 6; \quad (b)\ y = 6 + 9x - 4x^2$$

and ascertain if it be a maximum or a minimum in each case.

$(a)\quad \dfrac{dy}{dx} = 8x - 9 = 0;\ x = 1\tfrac{1}{8};\ \text{and}\ y = -11.0625$

$\dfrac{d^2y}{dx^2} = 8$; it is +; hence it is a minimum.

119

(b) $\dfrac{dy}{dx} = 9 - 8x = 0$; $x = 1\frac{1}{8}$; and $y = +11.0625$

$\dfrac{d^2y}{dx^2} = -8$; it is $-$; hence it is a maximum.

(2) Find the maxima and minima of the function

$$y = x^3 - 3x + 16$$

$$\frac{dy}{dx} = 3x^2 - 3 = 0; \; x^2 = 1; \text{ and } x = \pm 1$$

$$\frac{d^2y}{dx^2} = 6x; \text{ for } x = 1; \text{ it is } +,$$

hence $x = 1$ corresponds to a minimum $y = 14$. For $x = -1$ it is $-$; hence $x = -1$ corresponds to a maximum $y = +18$.

(3) Find the maxima and minima of $y = \dfrac{x-1}{x^2+2}$.

$$\frac{dy}{dx} = \frac{(x^2+2) \times 1 - (x-1) \times 2x}{(x^2+2)^2} = \frac{2x - x^2 + 2}{(x^2+2)^2} = 0$$

or $x^2 - 2x - 2 = 0$, whose solutions are $x = +2.73$ and $x = -0.73$.

$$\frac{d^2y}{dx^2} = -\frac{(x^2+2)^2(2-2x) - (2x - x^2 + 2)(4x^3 + 8x)}{(x^2+2)^4}$$

$$= \frac{2x^5 - 6x^4 - 8x^3 - 8x^2 - 24x + 8}{(x^2+2)^4}$$

The denominator is always positive, so it is sufficient to ascertain the sign of the numerator.

If we put $x = 2.73$, the numerator is negative; the maximum, $y = 0.183$.

If we put $x = -0.73$, the numerator is positive; the minimum, $y = -0.683$.

(4) The expense C of handling the products of a certain factory varies with the weekly output P according to the relation

$C = aP + \dfrac{b}{c + P} + d$, where a, b, c, d are positive constants. For what output will the expense be least?

$\dfrac{dC}{dP} = a - \dfrac{b}{(c + P)^2} = 0$ for maximum or minimum; hence

$a = \dfrac{b}{(c + P)^2}$ and $P = \pm\sqrt{\dfrac{b}{a}} - c$

As the output cannot be negative, $P = +\sqrt{\dfrac{b}{a}} - c$.

Now $\qquad \dfrac{d^2C}{dP^2} = +\dfrac{b(2c + 2P)}{(c + P)^4}$

which is positive for all the values of P; hence $P = +\sqrt{\dfrac{b}{a}} - c$ corresponds to a minimum.

(5) The total cost per hour C of lighting a building with N lamps of a certain kind is

$$C = N\left(\frac{C_l}{t} + \frac{EPC_e}{1000}\right)$$

where $\quad E$ is the commercial efficiency (watts per candle),

$\qquad P$ is the candle power of each lamp,

$\qquad t$ is the average life of each lamp in hours,

$\qquad C_l$ = cost of renewal in cents per hour of use,

$\qquad C_e$ = cost of energy per 1000 watts per hour.

Moreover, the relation connecting the average life of a lamp with the commercial efficiency at which it is run is approximately $t = mE^n$, where m and n are constants depending on the kind of lamp.

Find the commercial efficiency for which the total cost of lighting will be least.

We have $\qquad C = N\left(\dfrac{C_l}{m}E^{-n} + \dfrac{PC_e}{1000}E\right)$

$$\frac{dC}{dE} = N\left(\frac{PC_e}{1000} - \frac{nC_l}{m}E^{-(n+1)}\right) = 0$$

for maximum or minimum.

$$E^{n+1} = \frac{1000 \times nC_l}{mPC_e} \quad \text{and} \quad E = \sqrt[n+1]{\frac{1000 \times nC_l}{mPC_e}}$$

This is clearly for minimum, since

$$\frac{d^2C}{dE^2} = N\left[(n+1)\frac{nC_l}{m}E^{-(n+2)}\right]$$

which is positive for a positive value of E.

For a particular type of 16 candle-power lamps, $C_l = 17$ cents, $C_e = 5$ cents; and it was found that $m = 10$ and $n = 3.6$.

$$E = \sqrt[4.6]{\frac{1000 \times 3.6 \times 17}{10 \times 16 \times 5}} = 2.6 \text{ watts per candle power.}$$

Exercises X

(See answers on page 301)

You are advised to plot the graph of any numerical example.

(1) Find the maxima and minima of

$$y = x^3 + x^2 - 10x + 8$$

(2) Given $y = \frac{b}{a}x - cx^2$, find expressions for $\frac{dy}{dx}$, and for $\frac{d^2y}{dx^2}$; also find the value of x which makes y a maximum or a minimum, and show whether it is maximum or minimum.

(3) Find how many maxima and how many minima there are in the curve, the equation to which is

$$y = 1 - \frac{x^2}{2} + \frac{x^4}{24}$$

and how many in that of which the equation is

$$y = 1 - \frac{x^2}{2} + \frac{x^4}{24} - \frac{x^6}{720}$$

(4) Find the maxima and minima of

$$y = 2x + 1 + \frac{5}{x^2}$$

(5) Find the maxima and minima of

$$y = \frac{3}{x^2 + x + 1}$$

(6) Find the maxima and minima of

$$y = \frac{5x}{2 + x^2}$$

(7) Find the maxima and minima of

$$y = \frac{3x}{x^2 - 3} + \frac{x}{2} + 5$$

(8) Divide a number N into two parts in such a way that three times the square of one part plus twice the square of the other part shall be a minimum.

(9) The efficiency u of an electric generator at different values of output x is expressed by the general equation:

$$u = \frac{x}{a + bx + cx^2}$$

where a is a constant depending chiefly on the energy losses in the iron and c a constant depending chiefly on the resistance of the copper parts. Find an expression for that value of the output at which the efficiency will be a maximum.

(10) Suppose it to be known that consumption of coal by a certain steamer may be represented by the formula

$$y = 0.3 + 0.001v^3$$

where y is the number of tons of coal burned per hour and v is the speed expressed in nautical miles per hour. The cost of wages, interest on capital, and depreciation of that ship are together equal, per hour, to the cost of 1 ton of coal. What speed will make the total cost of a voyage of 1000 nautical miles a minimum? And, if coal costs 10 dollars per ton, what will that minimum cost of the voyage amount to?

(11) Find the maxima and minima of

$$y = \pm \frac{x}{6} \sqrt{x(10 - x)}$$

(12) Find the maxima and minima of

$$y = 4x^3 - x^2 - 2x + 1$$

CHAPTER 13

Partial Fraction and Inverse Functions

This chapter illustrates a technique that can be used to simplify differentiating a complex fraction. It is another tool for making things easier.

Partial Fractions

We have seen that when we differentiate a fraction we have to perform a rather complicated operation; and, if the fraction is not itself a simple one, the result is bound to be a complicated expression. If we could split the fraction into two or more simpler fractions such that their sum is equivalent to the original fraction, we could then proceed by differentiating each of these simpler expressions. And the result of differentiating would be the sum of two (or more) derivatives, each one of which is relatively simple; while the final expression, though of course it will be the same as that which could be obtained without resorting to this dodge, is thus obtained with much less effort and appears in a simplified form.

Let us see how to reach this result. Try first the job of adding two fractions together to form a resultant fraction. Take, for example, the two fractions $\dfrac{1}{x+1}$ and $\dfrac{2}{x-1}$. Every student can add these together and find their sum to be $\dfrac{3x+1}{x^2-1}$. And in the same way he can add together three or more fractions. Now this process can certainly be reversed: that is to say that, if this last expression were given, it is certain that it can somehow be split back again into its original components or partial fractions. Only we do not know in every case that may be presented to us *how* we can so split it. In order to find this out we shall consider a simple case at first. But it is important to bear in mind that all which follows applies only to what are called "proper" algebraic fractions, meaning fractions like the above, which have the numerator of *a lesser degree* than the denominator; that is, those in which the highest exponent of x is less in the numerator than in the denominator. If we have to deal with such an expression as $\dfrac{x^2+2}{x^2-1}$, we can simplify it by division, since it is equivalent to $1+\dfrac{3}{x^2-1}$; and $\dfrac{3}{x^2-1}$ is a proper algebraic fraction to which the operation of splitting into partial fractions can be applied, as explained hereafter.

Case I.   If we perform many additions of two or more fractions the denominators of which contain only terms in x, and no terms in x^2, x^3, or any other powers of x, we *always* find that *the denominator of the final resulting fraction is the product of the denominators* of the fractions which were added to form the result. It follows that by factorizing the denominator of this final fraction, we can find every one of the denominators of the partial fractions of which we are in search.

Suppose we wish to go back from $\dfrac{3x+1}{x^2-1}$ to the components which we know are $\dfrac{1}{x+1}$ and $\dfrac{2}{x-1}$. If we did not know what those components were we can still prepare the way by writing:

$$\frac{3x+1}{x^2-1}=\frac{3x+1}{(x+1)(x-1)}=\frac{}{x+1}+\frac{}{x-1}$$

126

leaving blank the places for the numerators until we know what to put there. We always may assume the sign between the partial fractions to be *plus,* since, if it be *minus,* we shall simply find the corresponding numerator to be negative. Now, since the partial fractions are *proper* fractions, the numerators are mere numbers without x at all, and we can call them $A, B, C \ldots$ as we please. So, in this case, we have:

$$\frac{3x + 1}{x^2 - 1} = \frac{A}{x + 1} + \frac{B}{x - 1}$$

If, now, we perform the addition of these two partial fractions, we get $\dfrac{A(x - 1) + B(x + 1)}{(x + 1)(x - 1)}$; and this must be equal to $\dfrac{3x + 1}{(x + 1)(x - 1)}$. And, as the denominators in these two expressions are the same, the numerators must be equal, giving us:

$$3x + 1 = A(x - 1) + B(x + 1)$$

Now, this is an equation with two unknown quantities, and it would seem that we need another equation before we can solve them and find A and B. But there is another way out of this difficulty. The equation must be true for all values of x; therefore it must be true for such values of x as will cause $x - 1$ and $x + 1$ to become zero, that is for $x = 1$ and for $x = -1$ respectively. If we make $x = 1$, we get $4 = (A \times 0) + (B \times 2)$, so that $B = 2$; and if we make $x = -1$, we get

$$-2 = (A \times -2) + (B \times 0)$$

so that $A = 1$. Replacing the A and B of the partial fractions by these new values, we find them to become $\dfrac{1}{x + 1}$ and $\dfrac{2}{x - 1}$; and the thing is done.

As a further example, let us take the fraction

$$\frac{4x^2 + 2x - 14}{x^3 + 3x^2 - x - 3}$$

The denominator becomes zero when x is given the value 1; hence $x - 1$ is a factor of it, and obviously then the other factor

127

will be $x^2 + 4x + 3$; and this can again be decomposed into $(x + 1)(x + 3)$. So we may write the fraction thus:

$$\frac{4x^2 + 2x - 14}{x^3 + 3x^2 - x - 3} = \frac{A}{x + 1} + \frac{B}{x - 1} + \frac{C}{x + 3}$$

making three partial factors.

Proceeding as before, we find

$$4x^2 + 2x - 14 = A(x - 1)(x + 3) + B(x + 1)(x + 3) + C(x + 1)(x - 1)$$

Now, if we make $x = 1$, we get:

$$-8 = (A \times 0) + B(2 \times 4) + (C \times 0); \text{ that is, } B = -1$$

If $x = -1$, we get

$$-12 = A(-2 \times 2) + (B \times 0) + (C \times 0); \text{ whence } A = 3.$$

If $x = -3$, we get:

$$16 = (A \times 0) + (B \times 0) + C(-2 \times -4); \text{ whence } C = 2.$$

So then the partial fractions are:

$$\frac{3}{x + 1} - \frac{1}{x - 1} + \frac{2}{x + 3}$$

which is far easier to differentiate with respect to x than the complicated expression from which it is derived.

Case II.   If some of the factors of the denominator contain terms in x^2, and are not conveniently put into factors, then the corresponding numerator may contain a term in x, as well as a simple number, and hence it becomes necessary to represent this unknown numerator not by the symbol A but by $Ax + B$; the rest of the calculation being made as before.

Try, for instance: $\dfrac{-x^2 - 3}{(x^2 + 1)(x + 1)}$

$$\frac{-x^2 - 3}{(x^2 + 1)(x + 1)} = \frac{Ax + B}{x^2 + 1} + \frac{C}{x + 1}$$

$$-x^2 - 3 = (Ax + B)(x + 1) + C(x^2 + 1)$$

Putting $x = -1$, we get $-4 = C \times 2$; and $C = -2$

hence $\qquad -x^2 - 3 = (Ax + B)(x + 1) - 2x^2 - 2$

and $\qquad x^2 - 1 = Ax(x + 1) + B(x + 1)$

Putting $x = 0$, we get $-1 = B$;

hence $\quad x^2 - 1 = Ax(x + 1) - x - 1$; or $\quad x^2 + x = Ax(x + 1)$

and $\qquad\qquad\qquad x + 1 = A(x + 1)$

so that $A = 1$, and the partial fractions are:

$$\frac{x - 1}{x^2 + 1} - \frac{2}{x + 1}$$

Take as another example the fraction

$$\frac{x^3 - 2}{(x^2 + 1)(x^2 + 2)}$$

We get $\quad \dfrac{x^3 - 2}{(x^2 + 1)(x^2 + 2)} = \dfrac{Ax + B}{x^2 + 1} + \dfrac{Cx + D}{x^2 + 2}$

$$= \frac{(Ax + B)(x^2 + 2) + (Cx + D)(x^2 + 1)}{(x^2 + 1)(x^2 + 2)}$$

In this case the determination of A, B, C, D is not so easy. It will be simpler to proceed as follows: Since the given fraction and the fraction found by adding the partial fractions are equal, and have *identical* denominators, the numerators must also be identically the same. In such a case, and for such algebraical expressions as those with which we are dealing here, *the coefficients of the same powers of x are equal and of same sign.*

Hence, since

$$x^3 - 2 = (Ax + B)(x^2 + 2) + (Cx + D)(x^2 + 1)$$

$$= (A + C)x^3 + (B + D)x^2 + (2A + C)x + 2B + D$$

we have $1 = A + C$; $0 = B + D$ (the coefficient of x^2 in the left expression being zero); $0 = 2A + C$; and $-2 = 2B + D$. Here are

129

four equations, from which we readily obtain $A = -1$; $B = -2$; $C = 2$; $D = 2$; so that the partial fractions are $\dfrac{2(x+1)}{x^2+2} - \dfrac{x+2}{x^2+1}$. This method can always be used; but the method shown first will be found the quickest in the case of factors in x only.

Case III. When among the factors of the denominator there are some which are raised to some power, one must allow for the possible existence of partial fractions having for denominator the several powers of that factor up to the highest. For instance, in splitting the fraction $\dfrac{3x^2 - 2x + 1}{(x+1)^2(x-2)}$ we must allow for the possible existence of a denominator $x + 1$ as well as $(x+1)^2$ and $x - 2$.

It may be thought, however, that, since the numerator of the fraction the denominator of which is $(x+1)^2$ may contain terms in x, we must allow for this in writing $Ax + B$ for its numerator, so that

$$\frac{3x^2 - 2x + 1}{(x+1)^2(x-2)} = \frac{Ax+B}{(x+1)^2} + \frac{C}{x+1} + \frac{D}{x-2}$$

If, however, we try to find A, B, C and D in this case, we fail, because we get four unknowns; and we have only three relations connecting them, yet

$$\frac{3x^2 - 2x + 1}{(x+1)^2(x-2)} = \frac{x-1}{(x+1)^2} + \frac{1}{x+1} + \frac{1}{x-2}$$

But if we write

$$\frac{3x^2 - 2x + 1}{(x+1)^2(x-2)} = \frac{A}{(x+1)^2} + \frac{B}{x+1} + \frac{C}{x-2}$$

we get $\quad 3x^2 - 2x + 1 = A(x-2) + B(x+1)(x-2) + C(x+1)^2.$

For $\quad x = -1; \quad 6 = -3A, \quad$ or $A = -2$

For $\quad x = 2; \quad 9 = 9C, \quad$ or $C = 1$

For $x = 0$; $1 = -2A - 2B + C$

Putting in the values of A and C:

$$1 = 4 - 2B + 1, \text{ from which } B = 2$$

Hence the partial fractions are:

$$\frac{2}{x+1} - \frac{2}{(x+1)^2} + \frac{1}{x-2}$$

instead of $\dfrac{1}{x+1} + \dfrac{x-1}{(x+1)^2} + \dfrac{1}{x-2}$ stated above as being the

fractions from which $\dfrac{3x^2 - 2x + 1}{(x+1)^2(x-2)}$ was obtained. The mystery

is cleared if we observe that $\dfrac{x-1}{(x+1)^2}$ can itself be split into the

two fractions $\dfrac{1}{x+1} - \dfrac{2}{(x+1)^2}$, so that the three fractions given

are really equivalent to

$$\frac{1}{x+1} + \frac{1}{x+1} - \frac{2}{(x+1)^2} + \frac{1}{x-2} = \frac{2}{x+1} - \frac{2}{(x+1)^2} + \frac{1}{x-2}$$

which are the partial fractions obtained.

We see that it is sufficient to allow for one numerical term in each numerator, and that we always get the ultimate partial fractions.

When there is a power of a factor of x^2 in the denominator, however, the corresponding numerators must be of the form $Ax + B$; for example,

$$\frac{3x-1}{(2x^2-1)^2(x+1)} = \frac{Ax+B}{(2x^2-1)^2} + \frac{Cx+D}{2x^2-1} + \frac{E}{x+1}$$

which gives

$$3x - 1 = (Ax+B)(x+1) + (Cx+D)(x+1)(2x^2-1) + E(2x^2-1)^2.$$

For $x = -1$, this gives $E = -4$. Replacing, transposing, collecting like terms, and dividing by $x + 1$, we get

$$16x^3 - 16x^2 + 3 = 2Cx^3 + 2Dx^2 + x(A - C) + (B - D).$$

Hence $2C = 16$ and $C = 8$; $2D = -16$ and $D = -8$; $A - C = 0$ or $A - 8 = 0$ and $A = 8$; and finally, $B - D = 3$ or $B = -5$. So that we obtain as the partial fractions:

$$\frac{8x - 5}{(2x^2 - 1)^2} + \frac{8(x - 1)}{2x^2 - 1} - \frac{4}{x + 1}$$

It is useful to check the results obtained. The simplest way is to replace x by a single value, say $+1$, both in the given expression and in the partial fractions obtained.

Whenever the denominator contains but a power of a single factor, a very quick method is as follows:

Taking, for example, $\dfrac{4x + 1}{(x + 1)^3}$, let $x + 1 = z$; then $x = z - 1$.

Replacing, we get

$$\frac{4(z - 1) + 1}{z^3} = \frac{4z - 3}{z^3} = \frac{4}{z^2} - \frac{3}{z^3}$$

The partial fractions are, therefore,

$$\frac{4}{(x + 1)^2} - \frac{3}{(x + 1)^3}$$

Applying this to differentiation, let it be required to differentiate $y = \dfrac{5 - 4x}{6x^2 + 7x - 3}$; we have

$$\frac{dy}{dx} = -\frac{(6x^2 + 7x - 3) \times 4 + (5 - 4x)(12x + 7)}{(6x^2 + 7x - 3)^2}$$

$$= \frac{24x^2 - 60x - 23}{(6x^2 + 7x - 3)^2}$$

If we split the given expression into

$$\frac{1}{3x-1} - \frac{2}{2x+3}$$

we get, however,

$$\frac{dy}{dx} = -\frac{3}{(3x-1)^2} + \frac{4}{(2x+3)^2}$$

which is really the same result as above split into partial fractions. But the splitting, if done after differentiating, is more complicated, as will easily be seen. When we shall deal with the *integration* of such expressions, we shall find the splitting into partial fractions a precious auxiliary.

Additional Problem Solving Examples

 Find the derivative of y=arc sin 4x.

 We use the formula for differentiation of the sin^{-1} or arc sin function, which states:
$$\frac{d}{dx}\ \sin^{-1}u = \frac{1}{\sqrt{1-u^2}}.$$

Hence

$$\frac{dy}{dx} = \frac{1}{\sqrt{1-16x^2}}(4) = \frac{4}{\sqrt{1-16x^2}}$$

 Given: $y = arc\ \tan\dfrac{3}{x}$, find $\dfrac{dy}{dx}$.

 In this example, we use the formula:

$$\frac{d(arc\ \tan\ u)}{dx} = \frac{1}{1+u^2} \cdot \frac{du}{dx}.$$

For

$$y = arc\ \tan\frac{3}{x},\quad u = \frac{3}{x},\quad and\quad du = \frac{-3}{x^2}.$$

133

Therefore,

$$\frac{dy}{dx} = \frac{1\left(\dfrac{-3}{x^2}\right)}{1+\left(\dfrac{3}{x}\right)^2} = \frac{\dfrac{-3}{x^2}}{\dfrac{x^2+9}{x^2}} = \frac{-3}{x^2+9}.$$

Exercises XI

(See answers on page 302)

Split into partial fractions:

(1) $\dfrac{3x+5}{(x-3)(x+4)}$

(2) $\dfrac{3x-4}{(x-1)(x-2)}$

(3) $\dfrac{3x+5}{x^2+x-12}$

(4) $\dfrac{x+1}{x^2-7x+12}$

(5) $\dfrac{x-8}{(2x+3)(3x-2)}$

(6) $\dfrac{x^2-13x+26}{(x-2)(x-3)(x-4)}$

(7) $\dfrac{x^2-3x+1}{(x-1)(x+2)(x-3)}$

(8) $\dfrac{5x^2+7x+1}{(2x+1)(3x-2)(3x+1)}$

(9) $\dfrac{x^2}{x^3-1}$

(10) $\dfrac{x^4+1}{x^3+1}$

(11) $\dfrac{5x^2+6x+4}{(x+1)(x^2+x+1)}$

(12) $\dfrac{x}{(x-1)(x-2)^2}$

(13) $\dfrac{x}{(x^2-1)(x+1)}$

(14) $\dfrac{x+3}{(x+2)^2(x-1)}$

(15) $\dfrac{3x^2+2x+1}{(x+2)(x^2+x+1)^2}$

(16) $\dfrac{5x^2+8x-12}{(x+4)^3}$

(17) $\dfrac{7x^2+9x-1}{(3x-2)^4}$

(18) $\dfrac{x^2}{(x^3-8)(x-2)}$

Derivative of an Inverse Function

Consider the function $y = 3x$; it can be expressed in the form $x = \frac{y}{3}$; this latter form is called the *inverse function* to the one originally given.

If $y = 3x$, $\frac{dy}{dx} = 3$; if $x = \frac{y}{3}$, $\frac{dx}{dy} = \frac{1}{3}$, and we see that

$$\frac{dy}{dx} = \frac{1}{\dfrac{dx}{dy}} \quad \text{or} \quad \frac{dy}{dx} \times \frac{dx}{dy} = 1$$

Consider $y = 4x^2$, $\frac{dy}{dx} = 8x$; the inverse function is

$$x = \frac{y^{\frac{1}{2}}}{2}, \quad \text{and} \quad \frac{dx}{dy} = \frac{1}{4\sqrt{y}} = \frac{1}{4 \times 2x} = \frac{1}{8x}$$

Here again $\qquad \frac{dy}{dx} \times \frac{dx}{dy} = 1$

It can be shown that for all functions which can be put into the inverse form, one can always write

$$\frac{dy}{dx} \times \frac{dx}{dy} = 1 \quad \text{or} \quad \frac{dy}{dx} = \frac{1}{\dfrac{dx}{dy}}$$

It follows that, being given a function, if it be easier to differentiate the inverse function, this may be done, and the reciprocal of the derivative of the inverse function gives the derivative of the given function itself.

As an example, suppose that we wish to differentiate $y = \sqrt{\dfrac{3}{x} - 1}$. We have seen one way of doing this, by writing

135

$u = \dfrac{3}{x} - 1$, and finding $\dfrac{dy}{du}$ and $\dfrac{du}{dx}$. This gives

$$\frac{dy}{dx} = -\frac{3}{2x^2\sqrt{\dfrac{3}{x}-1}}$$

If we had forgotten how to proceed by this method, or wished to check our result by some other way of obtaining the derivative, or for any other reason we could not use the ordinary method, we could proceed as follows: The inverse function is $x = \dfrac{3}{1+y^2}$.

$$\frac{dx}{dy} = -\frac{3 \times 2y}{(1+y^2)^2} = -\frac{6y}{(1+y^2)^2}$$

hence

$$\frac{dy}{dx} = \frac{1}{\dfrac{dx}{dy}} = -\frac{(1+y^2)^2}{6y} = -\frac{\left(1+\dfrac{3}{x}-1\right)^2}{6 \times \sqrt{\dfrac{3}{x}-1}} = -\frac{3}{2x^2\sqrt{\dfrac{3}{x}-1}}$$

Let us take, as another example, $y = \dfrac{1}{\sqrt[3]{\theta+5}}$

The inverse function $\theta = \dfrac{1}{y^3} - 5$ or $\theta = y^{-3} - 5$, and

$$\frac{d\theta}{dy} = -3y^{-4} = -3\sqrt[3]{(\theta+5)^4}$$

It follows that $\dfrac{dy}{d\theta} = -\dfrac{1}{3\sqrt[3]{(\theta+5)^4}}$, as might have been found otherwise.

We shall find this dodge most useful later on; meanwhile you are advised to become familiar with it by verifying by its means the results obtained in Exercises I (Chapter 4), Nos. 5, 6, 7; Examples (Chapter 9), Nos. 1, 2, 4; and Exercises VI (Chapter 9), Nos. 1, 2, 3 and 4.

You will surely realize from this chapter and the preceding, that in many respects the calculus is an *art* rather than a *science:* an art only to be acquired, as all other arts are, by practice. Hence you should work many examples, and set yourself other examples, to see if you can work them out, until the various artifices become familiar by use.

CHAPTER 14

On True Compound Interest and the Law of Organic Growth

Let there be a quantity growing in such a way that the increment of its growth, during a given time, shall always be proportional to its own magnitude. This resembles the process of reckoning interest on money at some fixed rate; for the bigger the capital, the bigger the amount of interest on it in a given time.

Now we must distinguish clearly between two cases, in our calculation, according as the calculation is made by what the arithmetic books call "simple interest", or by what they call "compound interest". For in the former case the capital remains fixed, while in the latter the interest is added to the capital, which therefore increases by successive additions.

(1) *At simple interest.* Consider a concrete case. Let the capital at start be $100, and let the rate of interest be 10 percent per annum. Then the increment to the owner of the capital will be $10 every year. Let him go on drawing his interest every year, and hoard it by putting it by in a stocking or locking it up in his safe. Then, if he goes on for 10 years, by the end of that time he will have received 10 increments of $10 each, or $100, making, with the original $100, a total of $200 in all. His property will have doubled itself in 10 years. If the rate of interest had been 5 percent, he would have had to hoard for 20 years to double his prop-

erty. If it had been only 2 percent, he would have had to hoard for 50 years. It is easy to see that if the value of the yearly interest is $\frac{1}{n}$ of the capital, he must go on hoarding for n years in order to double his property.

Or, if y be the original capital, and the yearly interest is $\frac{y}{n}$, then, at the end of n years, his property will be

$$y + n\frac{y}{n} = 2y$$

(2) *At compound interest.* As before, let the owner begin with a capital of \$100, earning interest at the rate of 10 percent per annum; but, instead of hoarding the interest, let it be added to the capital each year, so that the capital grows year by year. Then, at the end of one year, the capital will have grown to \$110; and in the second year (still at 10%) this will earn \$11 interest. He will start the third year with \$121, and the interest on that will be \$12.10; so that he starts the fourth year with \$133.10, and so on. It is easy to work it out, and find that at the end of the ten years the total capital will have grown to more than \$259. In fact, we see that at the end of each year, each dollar will have earned $\frac{1}{10}$ of a dollar, and therefore, if this is always added on, each year multiplies the capital by $\frac{11}{10}$; and if continued for ten years (which will multiply by this factor ten times over) the original capital will be multiplied by 2.59374. Let us put this into symbols. Put y_0 for the original capital; $\frac{1}{n}$ for the fraction added on at each of the n operations; and y_n for the value of the capital at the end of the nth operation. Then

$$y_n = y_0\left(1 + \frac{1}{n}\right)^n$$

But this mode of reckoning compound interest once a year is really not quite fair; for even during the first year the \$100 ought to have been growing. At the end of half a year it ought to have been at least \$105, and it certainly would have been fairer had the interest for the second half of the year been calculated on \$105. This would be equivalent to calling it 5% per half-year;

with 20 operations, therefore, at each of which the capital is multiplied by $\frac{21}{20}$. If reckoned this way, by the end of ten years the capital would have grown to more than \$265; for

$$\left(1 + \tfrac{1}{20}\right)^{20} = 2.653$$

But, even so, the process is still not quite fair; for, by the end of the first month, there will be some interest earned; and a half-yearly reckoning assumes that the capital remains stationary for six months at a time. Suppose we divided the year into 10 parts, and reckon a one percent interest for each tenth of the year. We now have 100 operations lasting over the ten years; or

$$y_n = \$100\left(1 + \tfrac{1}{100}\right)^{100}$$

which works out to \$270.48.

Even this is not final. Let the ten years be divided into 1000 periods, each of $\frac{1}{100}$ of a year; the interest being $\frac{1}{10}$ percent for each such period; then

$$y_n = \$100\left(1 + \tfrac{1}{1000}\right)^{1000}$$

which works out to a little more than \$271.69.

Go even more minutely, and divide the ten years into 10,000 parts, each $\frac{1}{1000}$ of a year, with interest at $\frac{1}{100}$ of 1 percent. Then

$$y_n = \$100\left(1 + \tfrac{1}{10,000}\right)^{10,000}$$

which amounts to about \$271.81.

Finally, it will be seen that what we are trying to find is in reality the ultimate value of the expression $\left(1 + \dfrac{1}{n}\right)^n$, which, as we see, is greater than 2; and which, as we take n larger and larger, grows closer and closer to a particular limiting value. However big you make n, the value of this expression grows nearer and nearer to the amount

$$2.71828 \ldots ,$$

a number *never to be forgotten.*

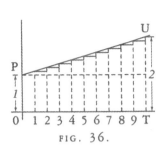

FIG. 36.

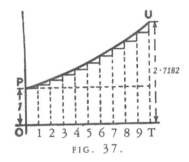

FIG. 37.

Let us take geometrical illustrations of these things. In Fig. 36, OP stands for the original value. OT is the whole time during which the value is growing. It is divided into 10 periods, in each of which there is an equal step up. Here $\dfrac{dy}{dx}$ is a constant; and if each step up is $\frac{1}{10}$ of the original OP, then, by 10 such steps, the height is doubled. If we had taken 20 steps, each of half the height shown, at the end the height would still be just doubled. Or n such steps, each of $\dfrac{1}{n}$ of the original height OP, would suffice to double the height. This is the case of simple interest. Here is 1 growing till it becomes 2.

In Fig. 37, we have the corresponding illustration of the geometrical progression. Each of the successive ordinates is to be $1 + \dfrac{1}{n}$, that is, $\dfrac{n+1}{n}$ times as high as its predecessor. The steps up are not equal, because each step up is now $\dfrac{1}{n}$ of the ordinate *at that part* of the curve. If we had literally 10 steps, with $\left(1 + \frac{1}{10}\right)$ for the multiplying factor, the final total would be $\left(1 + \frac{1}{10}\right)^{10}$ or 2.594 times the original 1. But if only we take n sufficiently large (and the corresponding $\dfrac{1}{n}$ sufficiently small), then the final value $\left(1 + \dfrac{1}{n}\right)^{n}$ to which unity will grow will be 2.71828. . . .

141

To this mysterious number 2.7182818..., the mathematicians have assigned the English letter *e*. All students know that the Greek letter π (called *pi*) stands for 3.141592..., but how many of them know that *e* represents 2.71828...? Yet it is an even more important number than π.

What, then, is *e*?

Suppose we were to let 1 grow at simple interest till it became 2; then, if at the same nominal rate of interest, and for the same time, we were to let 1 grow at true compound interest, instead of simple, it would grow to the value *e*.

This process of growing proportionately, at every instant, to the magnitude at that instant, some people call an *exponential rate* of growing. Unit exponential rate of growth is that rate which in unit time will cause 1 to grow to 2.718281. . . . It might also be called the *organic rate* of growing: because it is characteristic of organic growth (in certain circumstances) that the increment of the organism in a given time is proportional to the magnitude of the organism itself.

If we take 100 percent as the unit of rate, and any fixed period as the unit of time, then the result of letting 1 grow *arithmetically* at unit rate, for unit time, will be 2, while the result of letting 1 grow *exponentially* at unit rate, for the same time, will be 2.71828. . . .

A Little More About e. We have seen that we require to know what value is reached by the expression $\left(1 + \dfrac{1}{n}\right)^n$, when n becomes infinitely great. Arithmetically, here are tabulated a lot of values obtained by assuming $n = 2$; $n = 5$; $n = 10$; and so on, up to $n = 10,000$.

$$\left(1 + \tfrac{1}{2}\right)^2 \qquad = 2.25$$

$$\left(1 + \tfrac{1}{5}\right)^5 \qquad = 2.489$$

$$\left(1 + \tfrac{1}{10}\right)^{10} \qquad = 2.594$$

$$\left(1 + \tfrac{1}{20}\right)^{20} \qquad = 2.653$$

$$\left(1 + \tfrac{1}{100}\right)^{100} \qquad = 2.705$$

$$\left(1 + \tfrac{1}{1000}\right)^{1000} = 2.7169$$

$$\left(1 + \tfrac{1}{10,000}\right)^{10,000} = 2.7181$$

It is, however, worthwhile to find another way of calculating this immensely important figure.

Accordingly, we will avail ourselves of the binomial theorem, and expand the expression $\left(1 + \dfrac{1}{n}\right)^{n}$ in that well-known way.

The binomial theorem gives the rule that

$$(a + b)^n = a^n + n\,\frac{a^{n-1}b}{1!} + n(n-1)\frac{a^{n-2}b^2}{2!}$$

$$+ n(n-1)(n-2)\frac{a^{n-3}b^3}{3!} + \ldots.$$

Putting $a = 1$ and $b = \dfrac{1}{n}$, we get

$$\left(1 + \frac{1}{n}\right)^n = 1 + 1 + \frac{1}{2!}\left(\frac{n-1}{n}\right) + \frac{1}{3!}\frac{(n-1)(n-2)}{n^2}$$

$$+ \frac{1}{4!}\frac{(n-1)(n-2)(n-3)}{n^3} + \ldots.$$

Now, if we suppose n to become infinitely great, say a billion, or a billion billions, then $n - 1$, $n - 2$, and $n - 3$. etc., will all be sensibly equal to n; and then the series becomes

$$e = 1 + 1 + \frac{1}{2!} + \frac{1}{3!} + \frac{1}{4!} + \ldots.$$

By taking this rapidly convergent series to as many terms as we please, we can work out the sum to any desired point of accuracy. Here is the working for ten terms:

	1.000000
dividing by 1	1.000000
dividing by 2	0.500000

dividing by 3	0.166667
dividing by 4	0.041667
dividing by 5	0.008333
dividing by 6	0.001389
dividing by 7	0.000198
dividing by 8	0.000025
dividing by 9	<u>0.000003</u>
Total	<u>2.718282</u>

e is incommensurable with 1, and resembles in being an interminable non-recurrent decimal.

The Exponential Series. We shall have need of yet another series. Let us, again making use of the binomial theorem, expand the expression $\left(1 + \dfrac{1}{n}\right)^{nx}$, which is the same as e^x when we make n infinitely great.

$$e^x = 1^{nx} + nx \cdot \frac{1^{nx-1}\left(\dfrac{1}{n}\right)}{1!} + nx(nx-1)\frac{1^{nx-2}\left(\dfrac{1}{n}\right)^2}{2!}$$

$$+ nx(nx-1)(nx-2)\frac{1^{nx-3}\left(\dfrac{1}{n}\right)^3}{3!} + \ldots$$

$$= 1 + x + \frac{1}{2!} \cdot \frac{n^2x^2 - nx}{n^2} + \frac{1}{3!} \cdot \frac{n^3x^3 - 3n^2x^2 + 2nx}{n^3} + \ldots$$

$$= 1 + x + \frac{x^2 - \dfrac{x}{n}}{2!} + \frac{x^3 - \dfrac{3x^2}{n} + \dfrac{2x}{n^2}}{3!} + \ldots$$

But, when n is made infinitely great, this simplifies down to the following:

$$e^x = 1 + x + \frac{x^2}{2!} + \frac{x^3}{3!} + \frac{x^4}{4!} + \ldots$$

This series is called *the exponential series.*

The great reason why *e* is regarded of importance is that e^x possesses a property, not possessed by any other function of *x*, that *when you differentiate it its value remains unchanged;* or, in other words, its derivative is the same as itself. This can be instantly seen by differentiating it with respect to *x*, thus:

$$\frac{d(e^x)}{dx} = 0 + 1 + \frac{2x}{1 \cdot 2} + \frac{3x^2}{1 \cdot 2 \cdot 3} + \frac{4x^3}{1 \cdot 2 \cdot 3 \cdot 4} + \frac{5x^4}{1 \cdot 2 \cdot 3 \cdot 4 \cdot 5} + \ldots$$

or
$$= 1 + x + \frac{x^2}{1 \cdot 2} + \frac{x^3}{1 \cdot 2 \cdot 3} + \frac{x^4}{1 \cdot 2 \cdot 3 \cdot 4} + \ldots$$

which is exactly the same as the original series.

Now we might have gone to work the other way, and said: Go to; let us find a function of *x*, such that its derivative is the same as itself. Or, is there any expression, involving only powers of *x*, which is unchanged by differentiation? Accordingly, let us *assume* as a general expression that

$$y = A + Bx + Cx^2 + Dx^3 + Ex^4 + \ldots$$

(in which the coefficients *A, B, C,* . . . . will have to be determined), and differentiate it.

$$\frac{dy}{dx} = B + 2Cx + 3Dx^2 + 4Ex^3 + \ldots$$

Now, if this new expression is really to be the same as that from which it was derived, it is clear that *A must = B;* that $C = \frac{B}{2} = \frac{A}{1 \cdot 2}$; that $D = \frac{C}{3} = \frac{A}{1 \cdot 2 \cdot 3}$; that $E = \frac{D}{4} = \frac{A}{1 \cdot 2 \cdot 3 \cdot 4}$ . . . .

The law of change is therefore that

$$y = A\left(1 + \frac{x}{1} + \frac{x^2}{1 \cdot 2} + \frac{x^3}{1 \cdot 2 \cdot 3} + \frac{x^4}{1 \cdot 2 \cdot 3 \cdot 4} + \ldots\right)$$

If, now, we take A = 1 for the sake of further simplicity, we have

$$y = 1 + \frac{x}{1} + \frac{x^2}{1 \cdot 2} + \frac{x^3}{1 \cdot 2 \cdot 3} + \frac{x^4}{1 \cdot 2 \cdot 3 \cdot 4} + \ldots$$

Differentiating it any number of times will give always the same series over again.

If, now, we take the particular case of $A = 1$, and evaluate the series, we shall get simply

when $x = 1$, $y = 2.718281 \ldots$;   that is, $y = e$;

when $x = 2$, $y = (2.718281 \ldots)^2$;   that is, $y = e^2$;

when $x = 3$, $y = (2.718281 \ldots)^3$;   that is, $y = e^3$;

and therefore

when $x = x$, $y = (2.718281 \ldots)^x$;   that is, $y = e^x$,

thus finally demonstrating that

$$e^x = 1 + \frac{x}{1} + \frac{x^2}{1 \cdot 2} + \frac{x^3}{1 \cdot 2 \cdot 3} + \frac{x^4}{1 \cdot 2 \cdot 3 \cdot 4} + \ldots$$

Natural or Napierian Logarithms.

Another reason why e is important is because it was made by Napier, the inventor of logarithms, the basis of his system. If y is the value of e^x, then x is the *logarithm*, to the base e, of y. Or, if

$$y = e^x$$

then $x = \log_e y$

The two curves plotted in Figs. 38 and 39 represent these equations.

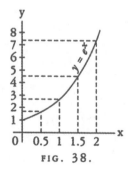

FIG. 38.

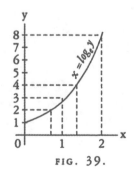

FIG. 39.

146

The points calculated are:

For Fig. 38

x	0	0.5	1	1.5	2
y	1	1.65	2.72	4.48	7.39

For Fig. 39

y	1	2	3	4	8
x	0	0.69	1.10	1.39	2.08

It will be seen that, though the calculations yield different points for plotting, yet the result is identical. The two equations really mean the same thing.

As many persons who use ordinary logarithms, which are calculated to base 10 instead of base e, are unfamiliar with the "natural" logarithms, it may be worth while to say a word about them. The ordinary rule that adding logarithms gives the logarithm of the product still holds good; or

$$\ln a + \ln b = \ln ab \quad \text{where } \ln = \log_e$$

Also the rule of powers holds good;

$$n \times \ln a = \ln a^n$$

But as 10 is no longer the basis, one cannot multiply by 100 or 1000 by merely adding 2 or 3 to the index. A natural logarithm is connected to the common logarithm of the same number by the relations:

$$\log_{10} x = \log_{10} e \times \ln x, \quad \text{and} \quad \ln x = \ln 10 \times \log_{10} x;$$

but $\quad \log_{10} e = \log_{10} 2.718 = 0.4343 \quad$ and $\quad \ln 10 = 2.3026$

$$\log_{10} x = 0.4343 \times \ln x$$

$$\ln x = 2.3026 \times \log_{10} x$$

A Useful Table of "Naperian Logarithms"

(Also called Natural Logarithms or Hyperbolic Logarithms)

Number	Log_e	Number	Log_e
1	0.0000	6	1.7918
1.1	0.0953	7	1.9459
1.2	0.1823	8	2.0794
1.5	0.4055	9	2.1972
1.7	0.5306	10	2.3026
2.0	0.6931	20	2.9957
2.2	0.7885	50	3.9120
2.5	0.9163	100	4.6052
2.7	0.9933	200	5.2983
2.8	1.0296	500	6.2146
3.0	1.0986	1,000	6.9078
3.5	1.2528	2,000	7.6009
4.0	1.3863	5,000	8.5172
4.5	1.5041	10,000	9.2103
5.0	1.6094	20,000	9.9035

Exponential and Logarithmic Equations.

Now let us try our hands at differentiating certain expressions that contain logarithms or exponentials.

Take the equation:

$$y = \ln x$$

First transform this into

$$e^y = x$$

whence, since the derivative of e^y with regard to y is the original function unchanged,

$$\frac{dx}{dy} = e^y$$

and, reverting from the inverse to the original function,

$$\frac{dy}{dx} = \frac{1}{\dfrac{dx}{dy}} = \frac{1}{e^y} = \frac{1}{x}$$

Now this is a very curious result. It may be written

$$\frac{d(\ln x)}{dx} = x^{-1}$$

Note that x^{-1} is a result that we could never have got by the rule for differentiating powers. That rule is to multiply by the power, and reduce the power by 1. Thus, differentiating x^3 gave us $3x^2$; and differentiating x^2 gave $2x^1$. But differentiating x^0 gives us $0 \times x^{-1} = 0$, because x^0 is itself $= 1$, and is a constant. We shall have to come back to this curious fact that differentiating $\log x$ gives us $\dfrac{1}{x}$ when we reach the chapter on integrating.

Now, try to differentiate

$$y = \ln(x + a)$$

that is

$$e^y = x + a$$

we have $\dfrac{d(x + a)}{dy} = e^y$, since the derivative of e^y remains e^y.

This gives

$$\frac{dx}{dy} = e^y = x + a;$$

hence, reverting to the original function, we get

$$\frac{dy}{dx} = \frac{1}{\dfrac{dx}{dy}} = \frac{1}{x + a}$$

Next try
$$y = \log_{10} x$$

First change to natural logarithms by multiplying by the modulus 0.4343. This gives us

$$y = 0.4343 \ln x$$

whence
$$\frac{dy}{dx} = \frac{0.4343}{x}$$

The next thing is not quite so simple. Try this:

$$y = a^x$$

Taking the logarithm of both sides, we get

$$\ln y = x \ln a$$

or
$$x = \frac{\ln y}{\ln a} = \frac{1}{\ln a} \times \ln y$$

Since $\dfrac{1}{\ln a}$ is a constant, we get

$$\frac{dx}{dy} = \frac{1}{\ln a} \times \frac{1}{y} = \frac{1}{a^x \times \ln a}$$

hence, reverting to the original function,

$$\frac{dy}{dx} = \frac{1}{\dfrac{dx}{dy}} = a^x \times \ln a$$

We see that, since

$$\frac{dx}{dy} \times \frac{dy}{dx} = 1 \quad \text{and} \quad \frac{dx}{dy} = \frac{1}{y} \times \frac{1}{\ln a}, \quad \frac{1}{y} \times \frac{dy}{dx} = \ln a$$

We shall find that whenever we have an expression such as

$\ln y = a$ function of x, we always have $\dfrac{1}{y}\dfrac{dy}{dx} =$ the derivative of the function of x, so that we could have written at once, from $\ln y = x \ln a$

$$\frac{1}{y}\frac{dy}{dx} = \ln a \quad \text{and} \quad \frac{dy}{dx} = y \ln a = a^x \ln a$$

Let us now attempt further examples.

Examples.
(1) $y = e^{-ax}$. Let $z = -ax$; then $y = e^z$

$$\frac{dy}{dz} = e^z; \frac{dz}{dx} = -a; \text{ hence } \frac{dy}{dx} = -ae^z = -ae^{-ax}$$

Or thus:

$$\ln y = -ax; \frac{1}{y}\frac{dy}{dx} = -a; \frac{dy}{dx} = -ay = -ae^{-ax}$$

(2) $y = e^{\frac{x^2}{3}}$. Let $z = \dfrac{x^2}{3}$; then $y = e^z$.

$$\frac{dy}{dz} = e^z; \frac{dz}{dx} = \frac{2x}{3}; \frac{dy}{dx} = \frac{2x}{3}e^{\frac{x^2}{3}}$$

Or thus: $\quad \ln y = \dfrac{x^2}{3}; \dfrac{1}{y}\dfrac{dy}{dx} = \dfrac{2x}{3}; \dfrac{dy}{dx} = \dfrac{2x}{3}e^{\frac{x^2}{3}}$

(3) $y = e^{\frac{2x}{x+1}}$. $\ln y = \dfrac{2x}{x+1}, \dfrac{1}{y}\dfrac{dy}{dx} = \dfrac{2(x+1) - 2x}{(x+1)^2}$

hence $\quad \dfrac{dy}{dx} = \dfrac{2y}{(x+1)^2} = \dfrac{2}{(x+1)^2}e^{\frac{2x}{x+1}}$

Check by writing $z = \dfrac{2x}{x+1}$

(4) $y = e^{\sqrt{x^2+a}}$. $\ln y = (x^2 + a)^{\frac{1}{2}}$

$$\frac{1}{y}\frac{dy}{dx} = \frac{x}{(x^2+a)^{\frac{1}{2}}} \quad \text{and} \quad \frac{dy}{dx} = \frac{x \times e^{\sqrt{x^2+a}}}{(x^2+a)^{\frac{1}{2}}}$$

For if $u = (x^2 + a)^{\frac{1}{2}}$ and $v = x^2 + a$, $u = v^{\frac{1}{2}}$,

$$\frac{du}{dv} = \frac{1}{2v^{\frac{1}{2}}}; \frac{dv}{dx} = 2x; \frac{du}{dx} = \frac{x}{(x^2+a)^{\frac{1}{2}}}$$

Check by writing $z = \sqrt{x^2 + a}$

(5) $y = \ln(a + x^3)$. Let $z = (a + x^3)$; then $y = \ln z$.

$$\frac{dy}{dz} = \frac{1}{z}; \frac{dz}{dx} = 3x^2; \text{ hence } \frac{dy}{dx} = \frac{3x^2}{a + x^3}$$

(6) $y = \ln\{3x^2 + \sqrt{a + x^2}\}$.   Let $z = 3x^2 + \sqrt{a + x^2}$;   then $y = \ln z$.

$$\frac{dy}{dz} = \frac{1}{z}; \frac{dz}{dx} = 6x + \frac{x}{\sqrt{x^2+a}}$$

$$\frac{dy}{dx} = \frac{6x + \dfrac{x}{\sqrt{x^2+a}}}{3x^2 + \sqrt{a+x^2}} = \frac{x\left(1 + 6\sqrt{x^2+a}\right)}{\left(3x^2 + \sqrt{x^2+a}\right)\sqrt{x^2+a}}$$

(7) $y = (x + 3)^2\sqrt{x - 2}$

$$\ln y = 2\ln(x + 3) + \tfrac{1}{2}\ln(x - 2)$$

$$\frac{1}{y}\frac{dy}{dx} = \frac{2}{(x+3)} + \frac{1}{2(x-2)}$$

$$\frac{dy}{dx} = (x+3)^2\sqrt{x-2}\left\{\frac{2}{x+3} + \frac{1}{2(x-2)}\right\} = \frac{5(x+3)(x-1)}{2\sqrt{x-2}}$$

(8) $y = (x^2 + 3)^3(x^3 - 2)^{\frac{2}{3}}$.

$$\ln y = 3\ln(x^2 + 3) + \tfrac{2}{3}\ln(x^3 - 2)$$

$$\frac{1}{y}\frac{dy}{dx} = 3\frac{2x}{x^2+3} + \frac{2}{3}\frac{3x^2}{x^3-2} = \frac{6x}{x^2+3} + \frac{2x^2}{x^3-2}$$

For if $u = \ln (x^2 + 3)$, $z = x^2 + 3$ and $u = \ln z$

$$\frac{du}{dz} = \frac{1}{z}; \frac{dz}{dx} = 2x; \frac{du}{dx} = \frac{2x}{z} = \frac{2x}{x^2 + 3}$$

Similarly, if $v = \ln (x^3 - 2)$, $\dfrac{dv}{dx} = \dfrac{3x^2}{x^3 - 2}$ and

$$\frac{dy}{dx} = (x^2 + 3)^3 (x^3 - 2)^{\frac{2}{3}} \left\{ \frac{6x}{x^2 + 3} + \frac{2x^2}{x^3 - 2} \right\}$$

(9) $y = \dfrac{\sqrt[2]{x^2 + a}}{\sqrt[3]{x^3 - a}}$

$$\ln y = \frac{1}{2} \ln (x^2 + a) - \frac{1}{3} \ln (x^3 - a)$$

$$\frac{1}{y} \frac{dy}{dx} = \frac{1}{2} \frac{2x}{x^2 + a} - \frac{1}{3} \frac{3x^2}{x^3 - a} = \frac{x}{x^2 + a} - \frac{x^2}{x^3 - a}$$

and $\quad \dfrac{dy}{dx} = \dfrac{\sqrt[2]{x^2 + a}}{\sqrt[3]{x^3 - a}} \left\{ \dfrac{x}{x^2 + a} - \dfrac{x^2}{x^3 - a} \right\}$

(10) $y = \dfrac{1}{\ln x}$.

$$\frac{dy}{dx} = \frac{\ln x \times 0 - 1 \times \dfrac{1}{x}}{\ln^2 x} = -\frac{1}{x \ln^2 x}$$

(11) $y = \sqrt[3]{\ln x} = (\ln x)^{\frac{1}{3}}$. Let $z = \ln x$; $y = z^{\frac{1}{3}}$,

$$\frac{dy}{dz} = \frac{1}{3} z^{-\frac{2}{3}}; \frac{dz}{dx} = \frac{1}{x}; \frac{dy}{dx} = \frac{1}{3x \sqrt[3]{\ln^2 x}}$$

(12) $y = \left(\dfrac{1}{a^x} \right)^{ax}$

153

$$\ln y = -ax \ln a^x = -ax^2 \cdot \ln a$$

$$\frac{1}{y}\frac{dy}{dx} = -2ax \cdot \ln a$$

and
$$\frac{dy}{dx} = -2ax\left(\frac{1}{a^x}\right)^{ax} \cdot \ln a = -2xa^{1-ax^2} \cdot \ln a$$

Try now the following exercises.

Exercises XII

(See answers on page 303)

(1) Differentiate $y = b(e^{ax} - e^{-ax})$.

(2) Find the derivative with respect to t of the expression $u = at^2 + 2 \ln t$.

(3) If $y = n^t$, find $\dfrac{d(\ln y)}{dt}$

(4) Show that if $y = \dfrac{1}{b} \cdot \dfrac{a^{bx}}{\ln a}; \dfrac{dy}{dx} = a^{bx}$

(5) If $w = pv^n$, find $\dfrac{dw}{dv}$

Differentiate

(6) $y = \ln x^n$ \qquad (7) $y = 3e^{-\frac{x}{x-1}}$ \qquad (8) $y = (3x^2 + 1)e^{-5x}$

(9) $y = \ln (x^a + a)$ \qquad (10) $y = (3x^2 - 1)(\sqrt{x} + 1)$

(11) $y = \dfrac{\ln (x + 3)}{x + 3}$ \qquad (12) $y = a^x \times x^a$

(13) It was shown by Lord Kelvin that the speed of signalling through a submarine cable depends on the value of the ratio of the external diameter of the core to the diameter of the enclosed copper wire. If this ratio is called y, then the number of signals s that can be sent per minute can be expressed by the formula

$$s = ay^2 \ln\frac{1}{y}$$

154

where a is a constant depending on the length and the quality of the materials. Show that if these are given, s will be a maximum if $1/e^{\frac{1}{2}}$.

(14) Find the maximum or minimum of

$$y = x^3 - \ln x$$

(15) Differentiate $y = \ln (axe^x)$

(16) Differentiate $y = (\ln ax)^3$

The Logarithmic Curve

Let us return to the curve which has its successive ordinates in geometrical progression, such as that represented by the equation $y = bp^x$.

We can see, by putting $x = 0$, that b is the initial height of y. Then when

$$x = 1, y = bp; \quad x = 2, y = bp^2; \quad x = 3, y = bp^3, \text{ etc.}$$

Also, we see that p is the numerical value of the ratio between the height of any ordinate and that of the next preceding it. In Fig. 40, we have taken p as $\frac{6}{5}$; each ordinate being $\frac{6}{5}$ as high as the preceding one.

If two successive ordinates are related together thus in a constant ratio, their logarithms will have a constant difference; so that, if we should plot out a new curve, Fig. 41, with values of ln y as ordinates, it would be a straight line sloping up by equal steps. In fact, it follows from the equation, that

$$\ln y = \ln b + x \cdot \ln p, \text{ whence } \ln y - \ln b = x \cdot \ln p.$$

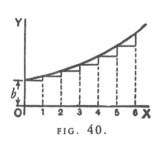

FIG. 40.

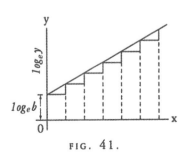

FIG. 41.

155

Now, since $\ln p$ is a mere number, and may be written as $\ln p = a$, it follows that

$$\ln \frac{y}{b} = ax$$

and the equation takes the new form

$$y = be^{ax}$$

The Die-away Curve

If we were to take p as a proper fraction (less than unity), the curve would obviously tend to sink downwards, as in Fig. 42, where each successive ordinate is $\frac{3}{4}$ of the height of the preceding one.

The equation is still

$$y = bp^x$$

but since p is less than one, $\ln p$ will be a negative quantity, and may be written $-a$; so that $p = e^{-a}$, and now our equation for the curve takes the form

$$y = be^{-ax}$$

The importance of this expression is that, in the case where the independent variable is *time,* the equation represents the course of a great many physical processes in which something is *gradually dying away.* Thus, the cooling of a hot body is represented (in Newton's celebrated "law of cooling") by the equation

$$\theta_t = \theta_0 e^{-at}$$

where θ_0 is the original excess of temperature of a hot body over that of its surroundings, θ_t the excess of temperature at the end of time t, and a is a constant—namely, the constant of decrement, depending on the amount of surface exposed by the body, and on its coefficients of conductivity and emissivity, etc.

A similar formula,

$$Q_t = Q_0 e^{-at}$$

is used to express the charge of an electrified body, originally having a charge Q_0, which is leaking away with a con

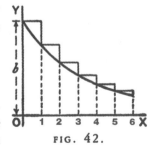

FIG. 42.

156

stant of decrement a; which constant depends in this case on the capacity of the body and on the resistance of the leakage-path.

Oscillations given to a flexible spring die out after a time, and the dying-out of the amplitude of the motion may be expressed in a similar way.

In fact e^{-at} serves as a *die-away factor* for all those phenomena in which the rate of decrease is proportional to the magnitude of that which is decreasing; or where, in our usual symbols, $\dfrac{dy}{dt}$ is proportional at every moment to the value that y has at that moment. For we have only to inspect the curve, Fig. 42, to see that at every part of it, the slope $\dfrac{dy}{dx}$ is proportional to the height y; the curve becoming flatter as y grows smaller. In symbols, thus

$$y = be^{-ax}$$

or
$$\ln y = \ln b - ax \ln e = \ln b - ax,$$

and, differentiating,
$$\frac{1}{y}\frac{dy}{dx} = -a$$

hence,
$$\frac{dy}{dx} = be^{-ax} \times (-a) = -ay$$

or, in words, the slope of the curve is downward, and proportional to y and to the constant a.

We should have got the same result if we had taken the equation in the form

$$y = bp^x$$

for then
$$\frac{dy}{dx} = bp^x \times \ln p$$

But
$$\ln p = -a$$

giving us
$$\frac{dy}{dx} = y \times (-a) = -ay$$

as before.

157

The Time-Constant. In the expression for the "die-away factor" e^{-at}, the quantity a is the reciprocal of another quantity known as "*the time-constant*", which we may denote by the symbol T. Then the die-away factor will be written $e^{-\frac{t}{T}}$; and it will be seen, by making $t = T$, that the meaning of T $\left(\text{or of } \dfrac{1}{a}\right)$ is that this is the length of time which it takes for the original quantity (called θ_0 or Q_0 in the preceding instances) to die away to $\dfrac{1}{e}th$ part—that is to 0.3679—of its original value.

The values of e^x and e^{-x} are continually required in different branches of physics, and as they are given in very few sets of mathematical tables, some of the values are tabulated here for convenience.

x	e^x	e^{-x}	$1 - e^{-x}$
0.00	1.0000	1.0000	0.0000
0.10	1.1052	0.9048	0.0952
0.20	1.2214	0.8187	0.1813
0.50	1.6487	0.6065	0.3935
0.75	2.1170	0.4724	0.5276
0.90	2.4596	0.4066	0.5934
1.00	2.7183	0.3679	0.6321
1.10	3.0042	0.3329	0.6671
1.20	3.3201	0.3012	0.6988
1.25	3.4903	0.2865	0.7135
1.50	4.4817	0.2231	0.7769
1.75	5.755	0.1738	0.8262
2.00	7.389	0.1353	0.8647
2.50	12.182	0.0821	0.9179
3.00	20.086	0.0498	0.9502
3.50	33.115	0.0302	0.9698
4.00	54.598	0.0183	0.9817
4.50	90.017	0.0111	0.9889
5.00	148.41	0.0067	0.9933
5.50	244.69	0.0041	0.9959
6.00	403.43	0.00248	0.99752
7.50	1808.04	0.00055	0.99945
10.00	22026.5	0.000045	0.999955

As an example of the use of this table, suppose there is a hot body cooling, and that at the beginning of the experiment (*i.e.* when $t = 0$) it is 72° hotter than the surrounding objects, and if the time-constant of its cooling is 20 minutes (that is, if it takes 20 minutes for its excess of temperature to fall to $\dfrac{1}{e}$ part of 72°), then we can calculate to what it will have fallen in any given time t. For instance, let t be 60 minutes. Then $\dfrac{t}{T} = 60 \div 20 = 3$, and we shall have to find the value of e^{-3}, and then multiply the original 72° by this. The table shows that e^{-3} is 0.0498. So that at the end of 60 minutes the excess of temperature will have fallen to $72° \times 0.0498 = 3.586°$.

Further Examples.
(1) The strength of an electric current in a conductor at a time t secs. after the application of the electromotive force producing it is given by the expression $C = \dfrac{E}{R}\left\{1 - e^{-\frac{Rt}{L}}\right\}$.

The time constant is $\dfrac{L}{R}$.

If $E = 10$, $R = 1$, $L = 0.01$; then when t is very large the term $1 - e^{-\frac{Rt}{L}}$ becomes 1, and $C = \dfrac{E}{R} = 10$; also

$$\frac{L}{R} = T = 0.01$$

Its value at any time may be written:
$$C = 10 - 10e^{-\frac{t}{0.01}}$$

the time-constant being 0.01. This means that it takes 0.01 sec. for the variable term to fall to $\dfrac{1}{e} = 0.3679$ of its initial value $10e^{-\frac{0}{0.01}} = 10$.

To find the value of the current when $t = 0.001$ sec., say, $\dfrac{t}{T} = 0.1$, $e^{-0.1} = 0.9048$ (from table).

It follows that, after 0.001 sec., the variable term is

$$0.9048 \times 10 = 9.048$$

and the actual current is $10 - 9.048 = 0.952$.

Similarly, at the end of 0.1 sec.,

$$\frac{t}{T} = 10; \ e^{-10} = 0.000045$$

the variable term is $10 \times 0.000045 = 0.00045$, the current being 9.9995.

(2) The intensity I of a beam of light which has passed through a thickness l cm. of some transparent medium is $I = I_0 e^{-Kl}$, where I_0 is the initial intensity of the beam and K is a "constant of absorption".

This constant is usually found by experiments. If it be found, for instance, that a beam of light has its intensity diminished by 18% in passing through 10 cm. of a certain transparent medium, this means that $82 = 100 \times e^{-K \times 10}$ or $e^{-10K} = 0.82$, and from the table one sees that $10K = 0.20$ very nearly; hence $K = 0.02$.

To find the thickness that will reduce the intensity to half its value, one must find the value of l which satisfies the equality $50 = 100 \times e^{-0.02l}$, or $0.5 = e^{-0.02l}$. It is found by putting this equation in its natural logarithmic form, namely,

$$l = \frac{\ln 0.5}{-0.02} = 34.7 \text{ cm. nearly.}$$

(3) The quantity Q of a radio-active substance which has not yet undergone transformation is known to be related to the initial quantity Q_0 of the substance by the relation $Q = Q_0 e^{-\lambda t}$, where λ is a constant and t the time in seconds elapsed since the transformation began.

For "Radium A", if time is expressed in seconds, experiment shows that $\lambda = 3.85 \times 10^{-3}$. Find the time required for transforming half the substance. (This time is called the "half life" of the substance.)

We have $\qquad 0.5 = e^{-0.00385t}$.

$$\log_{10} 0.5 = -0.00385t \times \log_{10} e$$

and $\qquad t = 3$ minutes very nearly.

Exercises XIII

(See answers on page 303)

(1) Draw the curve $y = be^{-\frac{t}{T}}$; where $b = 12$, $T = 8$, and t is given various values from 0 to 20.

(2) If a hot body cools so that in 24 minutes its excess of temperature has fallen to half the initial amount, deduce the time-constant, and find how long it will be in cooling down to 1 percent of the original excess.

(3) Plot the curve $y = 100(1 - e^{-2t})$.

(4) The following equations give very similar curves:

$$\text{(i) } y = \frac{ax}{x+b}; \quad \text{(ii) } y = a\left(1 - e^{-\frac{x}{b}}\right)$$

$$\text{(iii) } y = \frac{a}{90°} \arctan\left(\frac{x}{b}\right)$$

Draw all three curves, taking $a = 100$ millimetres; $b = 30$ millimetres.

(5) Find the derivative of y with respect to x, if

$$\text{(a) } y = x^x; \quad \text{(b) } y = (e^x)^x; \quad \text{(c) } y = e^{x^x}$$

(6) For "Thorium A", the value of λ is 5; find the "half life", that is, the time taken by the transformation of a quantity Q of "Thorium A" equal to half the initial quantity Q_0 in the expression

$$Q = Q_0 e^{-\lambda t};$$

t being in seconds.

(7) A condenser of capacity $K = 4 \times 10^{-6}$, charged to a potential $V_0 = 20$, is discharging through a resistance of 10,000 ohms. Find the potential V after (a) 0.1 second; (b) 0.01 second; assuming that the fall of potential follows the rule $V = V_0 e^{-\frac{t}{KR}}$.

161

(8) The charge Q of an electrified insulated metal sphere is reduced from 20 to 16 units in 10 minutes. Find the coefficient μ of leakage, if $Q = Q_0 \times e^{-\mu t}$; Q_0 being the initial charge and t being in seconds. Hence find the time taken by half the charge to leak away.

(9) The damping on a telephone line can be ascertained from the relation $i = i_0 e^{-\beta l}$, where i is the strength, after t seconds, of a telephonic current of initial strength i_0; l is the length of the line in kilometres, and β is a constant. For the Franco-English submarine cable laid in 1910, $\beta = 0.0114$. Find the damping at the end of the cable (40 kilometres), and the length along which i is still 8% of the original current (limiting value of very good audition).

(10) The pressure p of the atmosphere at an altitude h kilometres is approximately $p = p_0\, e^{-kh}$ (k is a constant); p_0 being the pressure at sea level (760 millimeters).

The pressures at 10, 20, and 50 kilometers being 199.2, 42.4, 0.32 millimeters respectively, find k in each case. Using the mean value of k, find the percentage error in each case.

(11) Find the minimum or maximum of $y = x^x$.

(12) Find the minimum or maximum of $y = x^{\frac{1}{x}}$.

(13) Find the minimum or maximum of $y = xa^{\frac{1}{x}}$.

CHAPTER 15

How to Deal With Sines And Cosines

Why We Study this Topic

Differentiating trigonometric functions differs somewhat from the functions we have operated on so far. In this chapter we learn the aspects that are unique to trigonometric functions.

Greek letters being usual to denote angles, we will take as the usual letter for any variable angle the letter θ ("theta").

Let us consider the function

$$y = \sin \theta$$

What we have to investigate is the value of $\dfrac{d(\sin \theta)}{d\theta}$; or, in other words, if the angle θ, we have to find the relation between the increment of the sine and the increment of the angle, both increments being infinitesimal small in themselves. Examine Fig. 43, wherein, if the radius of the circle is unity, the height of y is the sine, and θ is the angle. Now, if θ is supposed to increase by the addition to it of the small angle $d\theta$—an element of angle—the height of y, the sine, will be increased by a small element dy. The new height $y + dy$ will be the sine of the new angle $\theta + d\theta$, or, stating it as an equation,

$$y + dy = \sin (\theta + d\theta)$$

163

and subtracting from this the first equation gives

$$dy = \sin(\theta + d\theta) - \sin\theta$$

The quantity on the right-hand side is the difference between two sines, and books on trigonometry tell us how to work this out. For they tell us that if M and N are two different angles,

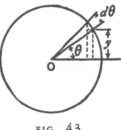

$$\sin M - \sin N = 2\cos\frac{M+N}{2}\cdot\sin\frac{M-N}{2}$$

If, then, we put $M = \theta + d\theta$ for one angle, and $N = \theta$ for the other, we may write

$$dy = 2\cos\frac{\theta + d\theta + \theta}{2}\cdot\sin\frac{\theta + d\theta - \theta}{2}$$

or,

$$dy = 2\cos(\theta + \tfrac{1}{2}d\theta)\cdot\sin\tfrac{1}{2}d\theta$$

But if we regard $d\theta$ as infinitesimal small, then in the limit we may neglect $\frac{1}{2}d\theta$ by comparison with θ, and may also take $\sin\frac{1}{2}d\theta$ as being the same as $\frac{1}{2}d\theta$. The equation then becomes:

$$dy = 2\cos\theta\cdot\tfrac{1}{2}d\theta$$

$$dy = \cos\theta\cdot d\theta$$

and, finally,

$$\frac{dy}{d\theta} = \cos\theta$$

The accompanying curves, Figs. 44 and 45, show, plotted to scale, the values of $y = \sin\theta$, and $\frac{dy}{d\theta} = \cos\theta$, for the corresponding values of θ.

Take next the cosine.

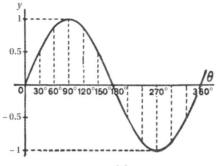

FIG. 44.

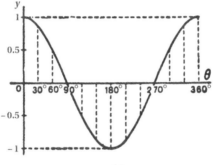

FIG. 45.

Let $y = \cos \theta$

Now $\cos \theta = \sin \left(\dfrac{\pi}{2} - \theta \right)$

Therefore

$$dy = d\left(\sin\left(\frac{\pi}{2} - \theta \right) \right) = \cos\left(\frac{\pi}{2} - \theta \right) \times d(-\theta)$$

$$= \cos\left(\frac{\pi}{2} - \theta \right) \times (-d\theta)$$

$$\frac{dy}{d\theta} = -\cos\left(\frac{\pi}{2} - \theta \right)$$

And it follows that

165

$$\frac{dy}{d\theta} = -\sin\theta$$

Lastly, take the tangent.

Let
$$y = \tan\theta$$
$$= \frac{\sin\theta}{\cos\theta}$$

Applying the rule given in Chapter 6 for differentiating a quotient of two functions, we get

$$\frac{dy}{d\theta} = \frac{\cos\theta\,\dfrac{d(\sin\theta)}{d\theta} - \sin\theta\,\dfrac{d(\cos\theta)}{d\theta}}{\cos^2\theta}$$

$$= \frac{\cos^2\theta + \sin^2\theta}{\cos^2\theta}$$

$$= \frac{1}{\cos^2\theta}$$

or
$$\frac{dy}{d\theta} = \sec^2\theta$$

Collecting these results, we have:

y	$\dfrac{dy}{d\theta}$
$\sin\theta$	$\cos\theta$
$\cos\theta$	$-\sin\theta$
$\tan\theta$	$\sec^2\theta$

Sometimes, in mechanical and physical questions, as, for example, in simple harmonic motion and in wave-motions, we have to deal with angles that increase in proportion to the time. Thus, if T be the time of one complete *period*, or movement round the circle, then, since the angle all round the circle is 2π radians, or $360°$, the amount of angle moved through in time t will be

$$\theta = 2\pi \frac{t}{T}, \text{ in radians}$$

or

$$\theta = 360\frac{t}{T}, \text{ in degrees}$$

If the *frequency*, or number of periods per second, be denoted by n, then $n = \frac{1}{T}$, and we may then write:

$$\theta = 2\pi n t$$

Then we shall have $\qquad y = \sin 2\pi n t$

If, now, we wish to know how the sine varies with respect to time, we must differentiate with respect, not to θ, but to t. For this we must resort to the artifice explained in Chapter 9, and put

$$\frac{dy}{dt} = \frac{dy}{d\theta} \cdot \frac{d\theta}{dt}$$

Now $\dfrac{d\theta}{dt}$ will obviously be $2\pi n$; so that

$$\frac{dy}{dt} = \cos \theta \times 2\pi n$$

$$= 2\pi n \cdot \cos 2\pi n t$$

Similarly, it follows that

$$\frac{d(\cos 2\pi n t)}{dt} = -2\pi n \cdot \sin 2\pi n t$$

Second Derivative of Sines or Cosines

We have seen that when $\sin \theta$ is differentiated with respect to θ it becomes $\cos \theta$; and that when $\cos \theta$ is differentiated with respect to θ it becomes $-\sin \theta$; or, in symbols,

$$\frac{d^2(\sin \theta)}{d\theta^2} = -\sin \theta$$

So we have this curious result that we have found a function such that if we differentiate it twice over, we get the same thing from which we started, but with the sign changed from $+$ to $-$.

The same thing is true for the cosine; for differentiating $\cos \theta$ gives us $-\sin \theta$, and differentiating $-\sin \theta$ gives us $-\cos \theta$; or thus:

$$\frac{d^2(\cos \theta)}{d\theta^2} = -\cos \theta.$$

Sines and cosines are functions of which the second derivative is equal and of opposite sign to the original function.

Examples.

With what we have so far learned we can now differentiate expressions of a more complex nature.

(1) $y = \arcsin x$.

If y is the angle whose sine is x, then $x = \sin y$.

$$\frac{dx}{dy} = \cos y$$

Passing now from the inverse function to the original one, we get

$$\frac{dy}{dx} = \frac{1}{\dfrac{dx}{dy}} = \frac{1}{\cos y}$$

Now $\qquad \cos y = \sqrt{1 - \sin^2 y} = \sqrt{1 - x^2}$

hence $\qquad \dfrac{dy}{dx} = \dfrac{1}{\sqrt{1 - x^2}}$

a rather unexpected result.

(2) $y = \cos^3 \theta$.

This is the same thing as $y = (\cos \theta)^3$

Let $v = \cos\ \theta$; then $y = v^3$; $\dfrac{dy}{dv} = 3v^2$

$$\frac{dv}{d\theta} = -\sin\ \theta$$

$$\frac{dy}{d\theta} = \frac{dy}{dv} \times \frac{dv}{d\theta} = -3\ \cos^2\theta\ \sin\ \theta$$

(3) $y = \sin\ (x + a)$.

Let $v = x + a$; then $y = \sin\ v$.

$$\frac{dv}{dx} = 1; \quad \frac{dy}{dv} = \cos v \quad \text{and} \quad \frac{dy}{dx} = \cos\ (x + a)$$

(4) $y = log_e \sin\ \theta$

Let $v = \sin\ \theta$; $y = log_e\ v$.

$$\frac{dv}{d\theta} = \cos\ \theta; \quad \frac{dy}{dv} = \frac{1}{v}; \quad \frac{dy}{d\theta} = \frac{1}{\sin\ \theta} \times \cos\ \theta = \cot\ \theta.$$

(5) $y = \cot\ \theta = \dfrac{\cos\ \theta}{\sin\ \theta}$

$$\frac{dy}{d\theta} = \frac{-\sin^2\theta - \cos^2\theta}{\sin^2\theta}$$

$$= -(1 + \cot^2\theta) = -\csc^2\theta$$

(6) $y = \tan\ 3\theta$

Let $v = 3\theta$; $y = \tan\ v$; $\dfrac{dv}{d\theta} = 3$; $\dfrac{dy}{dv} = \sec^2 v$

and
$$\frac{dy}{d\theta} = 3\ \sec^2 3\theta$$

(7) $y = \sqrt{1 + 3\ \tan^2\theta} = (1 + 3\ \tan^2\theta)^{\frac{1}{2}}$

Let $v = 3\ \tan^2\theta$, then $y = (1 + v)^{\frac{1}{2}}$;

Therefore $\dfrac{dv}{d\theta} = 6\ \tan\ \theta\ \sec^2\theta$; $\dfrac{dy}{dv} = \dfrac{1}{2\sqrt{1+v}}$

and
$$\frac{dy}{d\theta} = \frac{6 \tan \theta \sec^2\theta}{2\sqrt{1+v}} = \frac{6 \tan \theta \sec^2\theta}{2\sqrt{1+3 \tan^2\theta}}$$

for, if $u = \tan \theta$, $v = 3u^2$

$$\frac{du}{d\theta} = \sec^2\theta; \quad \frac{dv}{du} = 6u$$

hence
$$\frac{dv}{d\theta} = 6 \tan \theta \sec^2\theta$$

hence
$$\frac{dy}{d\theta} = \frac{6 \tan \theta \sec^2\theta}{2\sqrt{1+3 \tan^2\theta}}$$

(8) $y = \sin x \cos x$.

$$\frac{dy}{dx} = \sin x(-\sin x) + \cos x \times \cos x$$

$$= \cos^2 x - \sin^2 x$$

Closely associated with *sines*, *cosines* and *tangents* are three other very useful functions. They are the hyperbolic sine, cosine and tangent, and are written *sinh*, *cosh*, *tanh*. These functions are defined as follows:

$$\sinh x = \tfrac{1}{2}(e^x - e^{-x}), \quad \cosh x = \tfrac{1}{2}(e^x + e^{-x}),$$

$$\tanh x = \frac{\sinh x}{\cosh x} = \frac{e^x - e^{-x}}{e^x + e^{-x}}.$$

Between $\sinh x$ and $\cosh x$, there is an important relation, for

$$\cosh^2 x - \sinh^2 x = \tfrac{1}{4}(e^x + e^{-x})^2 - \tfrac{1}{4}(e^x - e^{-x})^2$$

$$= \tfrac{1}{4}(e^{2x} + 2 + e^{-2x} - e^{2x} + 2 - e^{-2x}) = 1.$$

Now $\dfrac{d}{dx} (\sinh x) = \tfrac{1}{2}(e^x + e^{-x}) = \cosh x.$

$$\frac{d}{dx} (\cosh x) = \tfrac{1}{2}(e^x - e^{-x}) = \sinh x.$$

$$\frac{d}{dx}(\tanh x) = \frac{\cosh x \frac{d}{dx}(\sinh x) - \sinh x \frac{d}{dx}(\cosh x)}{\cosh^2 x}$$

$$= \frac{\cosh^2 x - \sinh^2 x}{\cosh^2 x} = \frac{1}{\cosh^2 x}$$

by the relation just proved.

Trigonometric Differentiation

The three most basic trigonometric derivatives are:

$$\frac{d}{dx}(\sin x) = \cos x,$$

$$\frac{d}{dx}(\cos x) = -\sin x,$$

$$\frac{d}{dx}(\tan x) = \sec^2 x.$$

Given any trigonometric function, it can be differentiated by applying these basics in combination with the general rules for differentiating algebraic expressions.

The following will be most useful if committed to memory:

$$D_x \sin u = \cos u\ D_x u$$
$$D_x \cos u = -\sin u\ D_x u$$
$$D_x \tan u = \sec^2 u\ D_x u$$
$$D_x \sec u = \tan u \sec u\ D_x u$$
$$D_x \cot u = -\csc^2 u\ D_x u$$
$$D_x \csc u = -\csc u \cot u\ D_x u$$

Additional Problem Solving Examples

Find the derivative of: $y = \sin ax^2$.

Applying the theorem for the derivative of the sine of a function,

$$\frac{dy}{dx} = \cos ax^2 \cdot \frac{d}{dx}(ax^2)$$

$$= 2 ax \cos ax^2.$$

Find the derivative of: y tan 3θ.

Let   u = 3θ.

Then, y = tan u, and

$$\frac{dy}{d\theta} = \frac{dy}{du} \cdot \frac{du}{d\theta}$$

$$\frac{du}{d\theta} = 3,$$

and $\dfrac{dy}{du} = \sec^2 u.$

Therefore,

$$\frac{dy}{d\theta} = \frac{dy}{du} \cdot \frac{du}{d\theta} = \sec^2 u \cdot 3 = 3\sec^2(3\theta).$$

Inverse Trigonometric Differentiation

Find the derivative of y=arc sin 4x.

We use the formula for differentiation of the sin⁻¹ or arc sin function, which states:

$$\frac{d}{dx}\sin^{-1}u = \frac{1}{\sqrt{1-u^2}}.$$

Hence

$$\frac{dy}{dx} = \frac{1}{\sqrt{1-16x^2}}(4) = \frac{4}{\sqrt{1-16x^2}}$$

Q Given: $y = \text{arc } \tan \dfrac{3}{x}$, find $\dfrac{dy}{dx}$.

A In this example, we use the formula:

$$\frac{d(\text{arc } \tan u)}{dx} = \frac{1}{1+u^2} \cdot \frac{du}{dx}.$$

For

$$y = \text{arc } \tan \frac{3}{x}, \quad u = \frac{3}{x}, \quad \text{and} \quad du = \frac{-3}{x^2}.$$

Therefore,

$$\frac{dy}{dx} = \frac{1\left(\dfrac{-3}{x^2}\right)}{1+\left(\dfrac{3}{x}\right)^2} = \frac{\dfrac{-3}{x^2}}{\dfrac{x^2+9}{x^2}} = \frac{-3}{x^2+9}.$$

Exercises XIV

(See answers on page 303)

(1) Differentiate the following:

$$\text{(i) } y = A \sin\left(\theta - \frac{\pi}{2}\right)$$

$$\text{(ii) } y = \sin^2\theta; \text{ and } y = \sin 2\theta$$

$$\text{(iii) } y = \sin^3\theta; \text{ and } y = \sin 3\theta$$

(2) Find the value of θ for which $\sin\theta \times \cos\theta$ is a maximum.

(3) Differentiate $y = \dfrac{1}{2\pi} \cos 2\pi nt$.

(4) If $y = \sin a^x$, find $\dfrac{dy}{dx}$. (5) Differentiate $y = \ln \cos x$.

(6) Differentiate $y = 18.2 \sin (x + 26°)$.

(7) Plot the curve $y = 100 \sin (\theta - 15°)$; and show that the slope of the curve at $\theta = 75°$ is half the maximum slope.

(8) If $y = \sin \theta \cdot \sin 2\theta$, find $\dfrac{dy}{d\theta}$.

(9) If $y = a \cdot \tan^m(\theta^n)$, find the derivative of y with respect to θ.

(10) Differentiate $y = e^x \sin^2 x$.

(11) Differentiate the three equations of Exercises XIII, No. 4, and compare their derivatives, as to whether they are equal, or nearly equal, for very small values of x, or for very large values of x, or for values of x in the neighborhood of $x = 30$.

(12) Differentiate the following:

    (i) $y = \sec x.$ (ii) $y = \arccos x.$

    (iii) $y = \arctan x.$ (iv) $y = \operatorname{arcsec} x.$

    (v) $y = \tan x \times \sqrt{3 \sec x}.$

(13) Differentiate $y = \sin (2\theta + 3)^{2.3}$.

(14) Differentiate $y = \theta^3 + 3 \sin (\theta + 3) - 3^{\sin\theta} - 3^\theta$.

(15) Find the maximum or minimum of $y = \theta \cos \theta$.

Partial Differentiation

Until now, we differentiated functions having one independent variable only, e.g., $y = f(x)$. These functions can be looked upon as being plotted in a plane on y–x coordinates.

When we deal with surfaces, solids, or points in space, we encounter functions having two or more independent variables such as $z = f(x, y)$, for example. Graphically such a function can be plotted on xyz coordinate axes as follows:

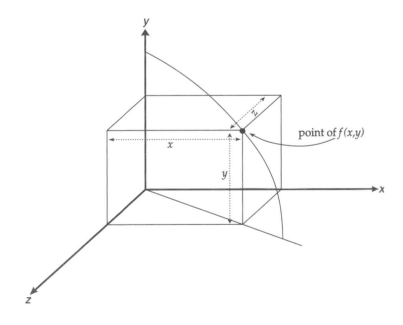

point of $f(x,y)$

We sometimes come across quantities that are functions of more than one independent variable. Thus, we may find a case where y depends on two other variable quantities, one of which we will call u and the other v. In symbols

$$y = f(u, v)$$

Take the simplest concrete case.

Let $\qquad\qquad\qquad y = u \times v$

What are we to do? If we were to treat v as a constant, and differentiate with respect to u, we should get

$$dy_v = v \, du$$

or if we treat u as a constant, and differentiate with respect to v, we should have:

$$dy_u = u \, dv$$

The little letters here put as subscripts are to show which quantity has been taken as constant in the operation.

Another way of indicating that the differentiation has been performed only *partially,* that is, has been performed only with respect to *one* of the independent variables, is to write the derivatives with Greek small deltas, instead of little d. In this way

$$\frac{\partial y}{\partial u} = v$$

$$\frac{\partial y}{\partial v} = u$$

If we put in these values for v and u respectively, we shall have

$$\left.\begin{aligned} dy_v &= \frac{\partial y}{\partial u} du, \\[2mm] dy_u &= \frac{\partial y}{\partial v} dv, \end{aligned}\right\} \quad \text{which are *partial derivatives.*}$$

176

But, if you think of it, you will observe that the total variation of y depends on *both* these things at the same time. That is to say, if both are varying, the real dy ought to be written

$$dy = \frac{\partial y}{\partial u}du + \frac{\partial y}{\partial v}dv$$

and this is called a *total differential.* In some books it is written
$$dy = \left(\frac{dy}{du}\right)du + \left(\frac{dy}{dv}\right)dv.$$

Example (1).   Find the partial derivatives of the expression $w = 2ax^2 + 3bxy + 4cy^3$. The answers are:

$$\left. \begin{aligned} \frac{\partial w}{\partial x} &= 4ax + 3by \\[2mm] \frac{\partial w}{\partial y} &= 3bx + 12cy^2 \end{aligned} \right\}$$

The first is obtained by supposing y constant, the second is obtained by supposing x constant; then the total differential is

$$dw = (4ax + 3by)dx + (3bx + 12cy^2)dy$$

Example (2).   Let $z = x^y$. Then, treating first y and then x as constant, we get in the usual way

$$\left. \begin{aligned} \frac{\partial z}{\partial x} &= yx^{y-1} \\[2mm] \frac{\partial z}{\partial y} &= x^y \times \log_e x \end{aligned} \right\}$$

so that $dz = yx^{y-1}dx + x^y \log_e x dy$.

Example (3).   A cone having height h and radius of base r, has volume $V = \frac{1}{3}\pi r^2 h$. If its height remains constant, while r changes, the ratio of change of volume, with respect to radius, is different from ratio of change of volume with respect to height which would occur if the height were varied and the radius kept constant, for

$$
\left.\begin{array}{l}
\dfrac{\partial V}{\partial r} = \dfrac{2\pi}{3} rh \\[3mm]
\dfrac{\partial V}{\partial h} = \dfrac{\pi}{3} r^2
\end{array}\right\}
$$

The variation when both the radius and the height change is given by $dV = \dfrac{2\pi}{3} rh\, dr + \dfrac{\pi}{3} r^2 dh$.

Example (4).   In the following example F and f denote two arbitrary functions of any form whatsoever. For example, they may be sine-functions, or exponentials, or mere algebraic functions of the two independent variables, t and x. This being understood, let us take the expression

$$y = F(x + at) + f(x - at)$$

or, $\qquad\qquad y = F(w) + f(v)$

where $\qquad\quad w = x + at, \quad\text{and}\quad v = x - at$

Then $\qquad \dfrac{\partial y}{\partial x} = \dfrac{\partial F(w)}{\partial w} \cdot \dfrac{\partial w}{\partial x} + \dfrac{\partial f(v)}{\partial v} \cdot \dfrac{\partial v}{\partial x}$

$$= F'(w) \cdot 1 + f'(v) \cdot 1$$

(where the figure 1 is simply the coefficient of x in w and v);

and $\qquad \dfrac{\partial^2 y}{\partial x^2} = F''(w) + f''(v)$

Also $\qquad \dfrac{\partial y}{\partial t} = \dfrac{\partial F(w)}{\partial w} \cdot \dfrac{\partial w}{\partial t} + \dfrac{\partial f(v)}{\partial v} \cdot \dfrac{\partial v}{\partial t}$

$$= F'(w) \cdot a - f'(v)a$$

and $\qquad \dfrac{\partial^2 y}{\partial t^2} = F''(w)a^2 + f''(v)a^2$

whence
$$\frac{\partial^2 y}{\partial t^2} = a^2 \frac{\partial^2 y}{\partial x^2}$$

This differential equation is of immense importance in mathematical physics.

Maxima and Minima of Functions of Two Independent Variables

Example (5).   Let us take up again Exercises IX, No. 4.

Let x and y be the lengths of two of the portions of the string. The third is $30 - (x + y)$, and the area of the triangle is $A = \sqrt{s(s - x)(s - y)(s - 30 + x + y)}$, where s is the half perimeter, so that $s = 15$, and $A = \sqrt{15P}$, where

$$P = (15 - x)(15 - y)(x + y - 15)$$

$$= xy^2 + x^2y - 15x^2 - 15y^2 - 45xy + 450x + 450y - 3375$$

Clearly A is a maximum when P is maximum.

$$dP = \frac{\partial P}{\partial x} dx + \frac{\partial P}{\partial y} dy$$

For a maximum (clearly it will not be a minimum in this case), one must have simultaneously

$$\frac{\partial P}{\partial x} = 0 \quad \text{and} \quad \frac{\partial P}{\partial y} = 0$$

that is,
$$\left. \begin{array}{l} 2xy - 30x + y^2 - 45y + 450 = 0, \\ 2xy - 30y + x^2 - 45x + 450 = 0. \end{array} \right\}$$

from which $(y - x)(y + x - 15) = 0$
since $y + x - 15 = 0$ does not yield a maximum, $x = y$.

If we now introduce this condition in the value of P, we find

$$P = (15 - x)^2(2x - 15) = 2x^3 - 75x^2 + 900x - 3375$$

For maximum or minimum, $\dfrac{dP}{dx} = 6x^2 - 150x + 900 = 0$, which gives $x = 15$ or $x = 10$.

Clearly $x = 15$ gives zero area; $x = 10$ gives the maximum, for $\dfrac{d^2P}{dx^2} = 12x - 150$, which is $+30$ for $x = 15$ and -30 for $x = 10$.

Example (6).   Find the dimensions of an ordinary railway coal truck with rectangular ends, so that, for a given volume V, the area of sides and floor together is as small as possible.

The truck is a rectangular box open at the top. Let x be the length and y be the width; then the depth is $\dfrac{V}{xy}$. The surface area is $S = xy + \dfrac{2V}{x} + \dfrac{2V}{y}$.

$$dS = \frac{\partial S}{\partial x}dx + \frac{\partial S}{\partial y}dy = \left(y - \frac{2V}{x^2}\right)dx + \left(x - \frac{2V}{y^2}\right)dy$$

For minimum (clearly it won't be a maximum here),

$$y - \frac{2V}{x^2} = 0 \quad x - \frac{2V}{y^2} = 0$$

From which $x = y$, $x^3 = 2V$ and $x = y = \sqrt[3]{2V}$.

Additional Problem Solving Examples

Let $u = u(x,y)$ be implicitly defined as a function of x and y by the equation $u + \ln u = xy$. Find

$$\frac{\partial u}{\partial x}, \frac{\partial u}{\partial y}, \frac{\partial^2 u}{\partial x \partial y} \text{ and } \frac{\partial^2 u}{\partial y \partial x}.$$

A   The partial derivative of a function is defined as follows: Let $z = f(x,y)$ be defined in a domain D of the xy-plane and let (x_1,y_1) be a point of D. Then $f(x,y_1)$ is a function depending only on x and if its derivative at the point x_1 exists, it is called the partial derivative of f with respect to x at (x_1,y_1) and is denoted by

$$\frac{\partial f}{\partial x}(x_1,y_1) \quad \text{or} \quad \frac{\partial z}{\partial x}\bigg|_{(x_1,y_1)}$$

If the point (x_1,y_1) is now allowed to vary, one obtains a new function of x and y (wherever the derivative exists) denoted by

$$\frac{\partial f}{\partial x}(x,y) = f_x(x,y) = \frac{\partial z(x,y)}{\partial x}.$$

Apparently, then, $\frac{\partial f}{\partial x}(x,y)$ may be obtained by simply treating y as a

constant and differentiating f with respect to its only remaining variable, x. In the case at hand, u is not defined explicitly as a function of x and y but the partial derivatives may be obtained by differentiating both sides of the defining equation with respect to x or y recalling that u depends on both x and y. Thus

$$\frac{\partial u(x,y)}{\partial x} + \frac{\partial l\,n[u(x,y)]}{\partial x} = \frac{\partial(xy)}{\partial x}. \tag{1}$$

Using the chain rule of single variable calculus, (1) becomes

$$\frac{\partial u}{\partial x} + \frac{d[l\,n\,u]}{dx}\frac{\partial u}{\partial x} = y \tag{2}$$

or $u_x + \frac{1}{u}u_x = y$. Therefore, $u_x\left(\frac{u+1}{u}\right) = y$, so that,

$$u_x = \frac{\partial u}{\partial x}\frac{uy}{u+1} = \frac{u(x,y)\cdot y}{u(x,y)+1}. \tag{3}$$

Also from the defining equation,

$$\frac{\partial u(x,y)}{\partial y} + \frac{\partial l\,n[u(x,y)]}{\partial y} = \frac{\partial(x,y)}{\partial y},$$

and the single variable chain rule can be used again to obtain

$$u_y + \frac{1}{u}u_y = x,$$

so that

181

$$u_y = \frac{\partial u}{\partial y} \frac{ux}{u+1} = \frac{u(x,y) \cdot x}{u(x,y)+1} \ . \tag{4}$$

The second partial derivative $\dfrac{\partial^2 u}{\partial x \partial y}$ is defined as $\dfrac{\partial (u_y)}{\partial x}$; that is, the partial derivative of the new function $\dfrac{\partial u}{\partial y}$ with respect to x. Thus, from (4), one obtains

$$\frac{\partial^2 u}{\partial x \partial y} = \frac{\partial}{\partial x}\left[\frac{ux}{u+1}\right] = \frac{u}{u+1} + x\frac{\partial}{\partial x}\left[\frac{u(x,y)}{u(x,y)+1}\right]$$

$$= \frac{u}{u+1} + x\frac{d}{du}\left[\frac{u}{u+1}\right] \cdot u_x$$

$$= \frac{u}{u+1} + \frac{x}{(u+1)^2} \cdot u_x = \frac{u}{u+1} + \frac{uxy}{(u+1)^3} \ . \tag{5}$$

Similarly,

$$\frac{\partial^2 u}{\partial y \partial x} = \frac{\partial}{\partial y}\left[\frac{uy}{u+1}\right] = \frac{u}{u+1} + y\frac{\partial}{\partial y}\left[\frac{u(x,y)}{u(x,y)+1}\right]$$

$$= \frac{u}{u+1} + y\frac{d}{du}\left[\frac{u}{u+1}\right] \cdot u_y$$

$$= \frac{u}{u+1} + \frac{y}{(u+1)^2} \cdot u_y = \frac{u}{u+1} + \frac{uxy}{(u+1)^3} \ . \tag{6}$$

It can be seen from (5) and (6) that $\dfrac{\partial^2 u}{\partial x \partial y} = \dfrac{\partial^2 u}{\partial y \partial x}$. This is a relation that can be proved to be true for all continuous functions $u(x,y)$.

 Find z_{xy} and z_{yx} from the expression:

$$z = x^2 y + 2xe^{\frac{1}{y}} \text{ and show that } z_{xy} = z_{yx}$$

 The first step of finding second partial derivatives is finding the first partial derivatives.

$$z_x = 2xy + 2e^{\frac{1}{y}}$$

and

$$z_y = x^2 - \frac{2xe^{\frac{1}{y}}}{y^2}.$$

To find z_{xy}, we differentiate z_y with respect to x.

$$z_{xy} = 2x - \frac{2e^{\frac{1}{y}}}{y^2}.$$

To find z_{yx} we differentiate z_x with respect to y, and obtain:

$$z_{yx} = 2x - \frac{2e^{\frac{1}{y}}}{y^2}.$$

Therefore $z_{yx} = z_{xy}$.

Exercises XV

(See answers on page 305)

(1) Differentiate the expression $\frac{x^3}{3} - 2x^3y - 2y^2x + \frac{y}{3}$ with respect to x alone, and with respect to y alone.

(2) Find the partial derivatives with respect to x, y, and z, of the expression

$$x^2yz + xy^2z + xyz^2 + x^2y^2z^2$$

(3) Let $r^2 = (x - a)^2 + (y - b)^2 + (z - c)^2$.

Find the value of $\frac{\partial r}{\partial x} + \frac{\partial r}{\partial y} + \frac{\partial r}{\partial z}$. Also find the value of $\frac{\partial^2 r}{\partial x^2} + \frac{\partial^2 r}{\partial y^2} + \frac{\partial^2 r}{\partial z^2}$.

(4) Find the total derivative of $y = u^v$.

(5) Find the total derivative of $y = u^3 \sin v$; of $y = (\sin x)^u$; and of $y = \frac{\ln u}{v}$.

(6) Verify that the sum of three quantities x, y, z, whose product is a constant k, is minimum when these three quantities are equal.

183

(7) Find the maximum or minimum of the function $u = x + 2xy + y$ if they exist.

(8) A post office regulation once stated that no parcel is to be of such a size that its length plus its girth exceeds 6 feet. What is the greatest volume that can be sent by post (*a*) in the case of a package of rectangular cross-section; (*b*) in the case of a package of circular cross-section?

(9) Divide π into 3 parts such that the product of their sines may be a maximum or minimum.

(10) Find the maximum or minimum of $u = \dfrac{e^{x+y}}{xy}$.

(11) Find maximum and minimum of

$$u = y + 2x - 2 \ln y - \ln x$$

(12) A bucket of given capacity has the shape of a horizontal isosceles triangular prism with the apex underneath, and the opposite face open. Find its dimensions in order that the least amount of iron sheet may be used in its construction.

CHAPTER 17

Integration

Why We Study this Topic

Review the introductory section, "Why We Study Calculus," where we describe how we obtain the areas of irregular shapes by summing up areas of a large number of narrow rectangles within small intervals. We also describe there, for example, how we obtain the volumes of irregular shaped bodies by summing up volumes of thin slices cut from the body. We noted, furthermore, that the thinner the slices, the greater is the accuracy in computing the volume of the body.

Through the use of integration the slices are made infinitesimally thin, so that the volume computation becomes precise. Integration performs the process of summing up the infinitesimally thin slices without having to add up laboriously a large number of thin slices. The integration process is carried out, moreover, rapidly and with relative ease.

The great secret has already been revealed that this mysterious symbol $\int$, which is after all only a long S, merely means "the sum of", or "the sum of all such quantities as". It therefore resembles that other symbol Σ (the Greek *Sigma*), which is also a sign of summation. There is this difference, however, in the practice of mathematicians as to the use of these signs, that while Σ is generally used to indicate the sum of a number of finite quantities, the integral sign $\int$ is generally used to indicate the summing up of a vast number of small quantities of infinitely minute magnitude, mere elements in fact, that go to make up the total required. Thus $\int dy = y$, and $\int dx = x$.

Anyone can understand how the whole of anything can be conceived of as made up of a lot of little bits; and the smaller the bits the more of them there will be. Thus, a line one inch long may be conceived as made up of 10 pieces, each $\frac{1}{10}$ of an inch long; or of 100 parts, each part being $\frac{1}{100}$ of an inch long; or of 1,000,000 parts, each of which is $\frac{1}{1,000,000}$ of an inch long; or, pushing the thought to the limits of conceivability, it may be regarded as made up of an infinite number of elements each of which is infinitesimally small.

Yes, you will say, but what is the use of thinking of anything that way? Why not think of it straight off, as a whole? The simple reason is that there are a vast number of cases in which one cannot calculate the bigness of the thing as a whole without reckoning up the sum of a lot of small parts. The process of *"integrating"* is to enable us to calculate totals that otherwise we should be unable to estimate directly.

Let us first take one or two simple cases to familiarize ourselves with this notion of summing up a lot of separate parts.

Consider the series:

$$1 + \tfrac{1}{2} + \tfrac{1}{4} + \tfrac{1}{8} + \tfrac{1}{16} + \tfrac{1}{32} + \tfrac{1}{64} + \ldots .$$

Here each member of the series is formed by taking half the value of the preceding one. What is the value of the total if we could go on to an infinite number of terms? Every schoolboy knows that the answer is 2. Think of it, if you like, as a line. Begin with one inch; add a half inch; add a quarter; add an eighth; and so on. If at any point of the operation we stop, there will still be a piece wanting to make up the whole 2 inches; and the piece wanting will always be the same size as the last piece added. Thus, if after having put together 1, $\frac{1}{2}$, and $\frac{1}{4}$, we stop, there will be $\frac{1}{4}$ wanting. If we go on till we have added $\frac{1}{64}$, there will still be

$\frac{1}{64}$ wanting. The remainder needed will always be equal to the last term added. By an infinite number of operations only should we reach the actual 2 inches. Practically we should reach it when we got to pieces so small that they could not be drawn—that would be after about 10 terms, for the eleventh term is $\frac{1}{1024}$. If we want to go so far that no measuring machine would detect it, we should merely have to go to about 20 terms. A microscope would not show even the 18th term! So the infinite number of operations is no such dreadful thing after all. The *integral* is simply the whole lot. But, as we shall see, there are cases in which the integral calculus enables us to get at the *exact* total that there would be as the result of an infinite number of operations. In such cases the integral calculus gives us a *rapid* and easy way of getting at a result

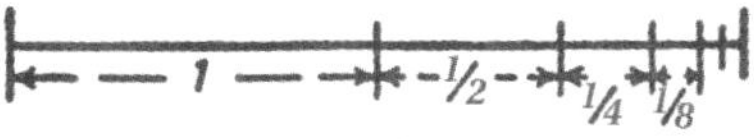

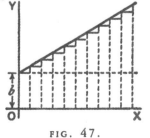

FIG. 46.

that would otherwise require an interminable lot of elaborate working out. So we had best lose no time in learning *how to integrate*.

Slopes of Curves and the Curves Themselves

Let us make a little preliminary enquiry about the slopes of curves. For we have seen that differentiating a curve means finding an expression for its slope (or for its slopes at different points). Can we perform the reverse process of reconstructing the whole curve if the slope (or slopes) are prescribed for us?

Go back to Chapter 10, case (2). Here we have the simplest of curves, a sloping line with the equation

$$y = ax + b$$

We know that here b represents the initial height of y when $x = 0$, and that a, which is the same as $\frac{dy}{dx}$, is the "slope" of the line. The line has a constant slope. All along it the elementary

FIG. 47.

187

triangles 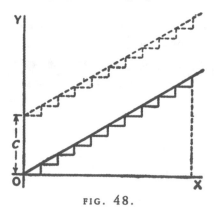 dy have the same proportion between height and base. Suppose we were to take the dx's and dy's of finite magnitude, so that 10 dx's made up one inch, then there would be ten little triangles like

◿ ◿ ◿ ◿ ◿ ◿ ◿ ◿ ◿ ◿

Now, suppose that we were ordered to reconstruct the "curve", starting merely from the information that $\dfrac{dy}{dx} = a$. What could we do? Still taking the little d's as of finite size, we could draw 10 of them, all with the same slope, and then put them together, end to end, like this:

And, as the slope is the same for all, they would join to make, as in Fig. 48, a sloping line sloping with the correct slope

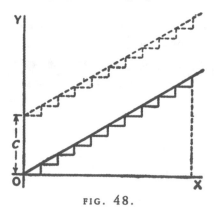

FIG. 48.

$\dfrac{dy}{dx} = a$. And whether we take the dy's and dx's as finite or infinitely small, as they are all alike, clearly $\dfrac{y}{x} = a$, if we reckon y as the total of all the dy's, and x as the total of all the dx's. But whereabouts are we to put this sloping line? Are we to start at the origin O, or higher up? As the only information we have is as to the slope, we are without any instructions as to the particular height above O; in fact the initial height is undetermined. The slope will be the same, whatever the initial height. Let us therefore make a shot at what may be wanted, and start the sloping line at a height C above O. That is, we have the equation

$$y = ax + C$$

It becomes evident now that in this case the added constant means the particular value that y has when $x = 0$.

Now let us take a harder case, that of a line, the slope of which is not constant, but turns up more and more. Let us assume that the upward slope gets greater and greater in proportion as x grows. In symbols this is:

$$\frac{dy}{dx} = ax$$

Or, to give a concrete case, take $a = \frac{1}{5}$, so that

$$\frac{dy}{dx} = \frac{1}{5}x$$

Then we had best begin by calculating a few of the values of the slope at different values of x, and also draw little diagrams of them.

When

$x = 0$, $\frac{dy}{dx} = 0$,

$x = 1$, $\frac{dy}{dx} = 0 \cdot 2$,

$x = 2$, $\frac{dy}{dx} = 0 \cdot 4$,

$x = 3$, $\frac{dy}{dx} = 0 \cdot 6$,

$x = 4$, $\frac{dy}{dx} = 0 \cdot 8$,

$x = 5$, $\frac{dy}{dx} = 1 \cdot 0$.

189

Now try to put the pieces together, setting each so that the middle of its base is the proper distance to the right, and so that they fit together at the corners; thus (Fig. 49). The result is, of course, not a smooth curve: but it is an approximation to one. If we had taken bits half as long, and twice as numerous, like Fig. 50, we should have a better approximation.

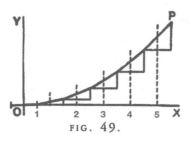

FIG. 49.

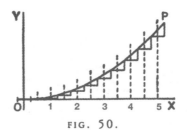

FIG. 50.

But for a perfect curve we ought to take each dx and its corresponding dy infinitesimally small, and infinitely numerous.

Then, how much ought the value of any y to be? Clearly, at any point P of the curve, the value of y will be the sum of all the little dy's from 0 up to that level, that is to say, $\int dy = y$. And as each dy is equal to $\frac{1}{5}x \cdot dx$, it follows that the whole y will be equal to the sum of all such bits as $\frac{1}{5}x \cdot dx$, or, as we should write it, $\int \frac{1}{5}x \cdot dx$.

Now if x had been constant, $\int \frac{1}{5}x \cdot dx$ would have been the same as $\frac{1}{5}x \int dx$, or $\frac{1}{5}x^2$. But x began by being 0, and increases to the particular value of x at the point P, so that its average value from 0 to that point is $\frac{1}{2}x$. Hence $\int \frac{1}{5}xdx = \frac{1}{10}x^2$; or $y = \frac{1}{10}x^2$.

But, as in the previous case, this requires the addition of an undetermined constant C, because we have not been told at what height above the origin the curve will begin, when $x = 0$. So we write, as the equation of the curve drawn in Fig. 51,

$$y = \frac{1}{10}x^2 + C$$

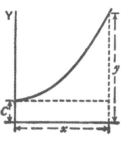

FIG. 51.

Exercises XVI

(See answers on page 305)

(1) Find the ultimate sum of $\frac{2}{3} + \frac{1}{3} + \frac{1}{6} + \frac{1}{12} + \frac{1}{24} + \ldots$

(2) Show that the series $1 - \frac{1}{2} + \frac{1}{3} - \frac{1}{4} + \frac{1}{5} - \frac{1}{6} + \frac{1}{7} \ldots$, is convergent, and find its sum to 8 terms.

(3) If $\ln(1 + x) = x - \dfrac{x^2}{2} + \dfrac{x^3}{3} - \dfrac{x^4}{4} + \ldots$, find $\ln 1.3$.

(4) Following a reasoning similar to that explained in this chapter, find y,

$$\text{if } \frac{dy}{dx} = \frac{1}{4}x$$

(5) If $\dfrac{dy}{dx} = 2x + 3$, find y

Additional Problem Solving Examples

 Determine the area under the curve: $y = f(x) = x^2$ between $x = 2$ and $x = 3$.

 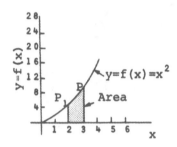

It is given that the area to be evaluated is between $x = 2$ and $x = 3$, therefore, these are the limits of the integral which gives us the required area. Area is equal to the integral of the upper function minus the lower function. From the diagram it is seen that the required area is between $y = x^2$ as the upper function and $y = 0$ (the x-axis) as the lower function. Therefore, we can write:

$$A = \int_2^3 (x^2 - 0)dx$$

$$= \int_2^3 x^2 dx$$

$$= \left. \frac{x^3}{3} \right]_2^3$$

$$A = \frac{3^3}{3} - \frac{2^3}{3} = \frac{19}{3} \; .$$

 Find the area between the curve: $y = x^3$, and the x-axis, from $x = -2$ to $x = 3$.

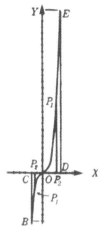

It is generally advantageous to sketch the curve, since parts of the curve may have to be considered separately, particularly when positive and negative limits are given. The desired area is composed of the two parts: BOC and ODE. To find the total area, we can evaluate each area separately and then add. The area is the integral of the upper function minus the lower function. In the first quadrant, the upper function is the curve $y = x^3$, the lower function is $y=0$, (the x-axis) and the limits are $x = 0$ and $x = 3$. In the third quadrant, the upper function is $y = 0$, the lower function is the curve $y = x^3$, and the limits $x = -2$ and $x = 0$. Hence we can write,

$$A_{total} = \int_0^3 (x^3 - 0)\, dx + \int_{-2}^0 (0 - x^3)\, dx$$

$$= \int_0^3 x^3 dx + \int_{-2}^0 -x^3 dx$$

$$= \left[\frac{x^4}{4} \right]_0^3 + \left[-\frac{x^4}{4} \right]_{-2}^0$$

$$= \frac{81}{4} + \frac{16}{4}$$

$$= 24\frac{1}{4} \quad \text{sq. units}$$

Refusal to consider this problem in two parts does <u>not</u> give area, but gives "net area" with one area considered positive and the other negative.

Integrate the expression: $\int \dfrac{dx}{1+e^x}$.

 We wish to convert the given integral into the form $\int \dfrac{du}{u}$. If we multiply $\dfrac{1}{1+e^x}$ by $\dfrac{e^{-x}}{e^{-x}}$ (which is equal to 1) we obtain:

$$\frac{e^{-x}(1)}{e^{-x}(1+e^x)} = \frac{e^{-x}}{e^{-x}+e^0} = \frac{e^{-x}}{e^{-x}+1} .$$

In integrating this, we apply the formula, $\int \dfrac{du}{u} = \ln \ |u| + C$, letting

$u = e^{-x} + 1$. Then $du = -e^{-x}dx$. We obtain:

$$\int \frac{e^{-x}}{e^{-x}+1} \ dx = -\int \frac{-e^{-x}\,dx}{e^{-x}+1} = -\ln(1+e^{-x}) + C.$$

 Integrate: $\int \dfrac{2x}{x+1}\, dx.$

 To integrate the given expression we manipulate the integrand to obtain the form $\int \dfrac{du}{u}$. This can be done as follows:

$$\int \frac{2x}{x+1} \ dx = 2 \int \frac{x}{x+1} \ dx$$

$$= 2 \int \left(\frac{x+1}{x+1} - \frac{1}{x+1} \right) dx$$

$$= 2 \int \left(1 - \frac{1}{x+1} \right) dx$$

$$= 2 \int dx - 2 \int \frac{dx}{x+1} \ .$$

Now, applying the formula $\int \dfrac{du}{u} = \ln u$, we obtain:

$$\int \frac{2x}{x+1} \ dx = 2x - 2 \ \ln(x+1) + C.$$

CHAPTER 18

Integration as the Reverse of Differentiating

How Is Integration Related to the Preceding Topic of Differentiation

In studying differentiation we found that we can, for example, obtain the speed of a vehicle from knowing the vehicle's position as a function of time. We may want to obtain, on the other hand, the vehicle's position from knowing the speed of the vehicle as a function of time. The vehicle's speed can be measured, for example, as the vehicle travels along its path. It is seen that obtaining the vehicle's position from its speed is the reverse situation from obtaining the vehicle's speed from its position. The latter situation can be realized by using differentiation, whereas the reverse situation of obtaining the vehicle's position from its speed can be realized by integration.

Integration can thus be regarded as the opposite of differentiation. For this reason, integration is often described as being an anti-derivative process.

Differentiation is a much simpler technique than integration. Differentiation can be carried out on even very complex functions with only handful of formulas that can be memorized. Integration, often referred to as anti-differentiation, on the other

195

hand, may require complex techniques for even simple functions, and long tables of integrals with numerous formulas have been assembled to help with the integration process.

Differentiating is the process by which when y is given us as a function of x, we can find $\dfrac{dy}{dx}$.

Like every other mathematical operation, the process of differentiation may be reversed. Thus, if differentiating $y = x^4$ gives us $\dfrac{dy}{dx} = 4x^3$, then, if one begins with $\dfrac{dy}{dx} = 4x^3$, one would say that reversing the process would yield $y = x^4$. But here comes in a curious point. We should get $\dfrac{dy}{dx} = 4x^3$ if we had begun with *any* of the following: x^4, or $x^4 + a$, or $x^4 + c$, or x^4 with *any* added constant. So it is clear that in working backwards from $\dfrac{dy}{dx}$ to y, one must make provision for the possibility of there being an added constant, the value of which will be undetermined until ascertained in some other way. So, if differentiating x^n yields nx^{n-1}, going backwards from $\dfrac{dy}{dx} = nx^{n-1}$ will give us $y = x^n + C$; where C stands for the yet undetermined possible constant.

Clearly, in dealing with powers of x, the rule for working backwards will be: Increase the power by 1, then divide by that increased power, and add the undetermined constant.

So, in the case where

$$\frac{dy}{dx} = x^n$$

working backwards, we get

$$y = \frac{1}{n+1}x^{n+1} + C$$

If differentiating the equation $y = ax^n$ gives us

$$\frac{dy}{dx} = anx^{n-1}$$

it is a matter of common sense that beginning with

$$\frac{dy}{dx} = anx^{n-1}$$

and reversing the process, will give us

$$y = ax^n$$

So, when we are dealing with a multiplying constant, we must simply put the constant as a multiplier of the result of the integration.

Thus, if $\frac{dy}{dx} = 4x^2$, the reverse process gives us $y = \frac{4}{3}x^3$.

But this is incomplete. For we must remember that if we had started with

$$y = ax^n + C$$

where C is any constant quantity whatever, we should equally have found

$$\frac{dy}{dx} = anx^{n-1}$$

So, therefore, when we reverse the process we must always remember to add on this undetermined constant, even if we do not yet know what its value will be.

This process, the reverse of differentiating, is called *integrating*; for it consists in finding the value of the whole quantity y when you are given only an expression for dy or for $\frac{dy}{dx}$. Hitherto we have as much as possible kept dy and dx together as a derivative: henceforth we shall more often have to separate them.

If we begin with a simple case,

$$\frac{dy}{dx} = x^2$$

We may write this, if we like, as

$$dy = x^2 dx$$

Now this is a "differential equation" which informs us that an element of y is equal to the corresponding element of x multiplied by x^2. Now, what we want is the integral; therefore, write down with the proper symbol the instructions to integrate both sides, thus:

$$\int dy = \int x^2 dx$$

[Note as to reading integrals: the above would be read thus:

"Integral of dee-wy equals *integral of eks-squared dee-eks."*]

We haven't yet integrated: we have only written down instructions to integrate—if we can. Let us try. Plenty of other fools can do it—why not we also? The left-hand side is simplicity itself.

The sum of all the bits of y is the same thing as y itself. So we may at once put:

$$y = \int x^2 dx$$

But when we come to the right-hand side of the equation we must remember that what we have got to sum up together is not all the dx's, but all such terms as $x^2 dx$; and this will *not* be the

same as $x^2 \int dx$, because x^2 is not a constant. For some of the dx's will be multiplied by big values of x^2, and some will be multiplied by small values of x^2, according to what x happens to be. So we must bethink ourselves as to what we know about this process of integration being the reverse of differentiation. Now, our rule for this reversed process when dealing with x'' is "increase the power by one, and divide by the same number as this increased power". That is to say, $x^2 dx$ will be changed* to $\frac{1}{3}x^3$. Put this into the equation; but don't forget to add the "constant of integration" C at the end. So we get:

$$y = \tfrac{1}{3}x^3 + C$$

You have actually performed the integration. How easy!
Let us try another simple case

Let $$\frac{dy}{dx} = ax^{12}$$

where a is any constant multiplier. Well, we found when differentiating (see Chapter 5) that any constant factor in the value of y reappeared unchanged in the value of $\dfrac{dy}{dx}$. In the reversed process of integrating, it will therefore also reappear in the value of y.

---

*You may ask: what has become of the little dx at the end? Well, remember that it was really part of the derivative, and when changed over to the right-hand side, as in the $x^2 dx$, serves as a reminder that x is the independent variable with respect to which the operation is to be effected; and, as the result of the product being totalled up, the power of x has increased by *one*. You will soon become familiar with all this.

So we may go to work as before, thus:

$$dy = ax^{12} \cdot dx$$

$$\int dy = \int ax^{12} \cdot dx$$

$$\int dy = a \int x^{12} dx$$

$$y = a \times \tfrac{1}{13} x^{13} + C$$

So that is done. How easy!

We begin to realize now that integrating is a process *of finding our way back,* as compared with differentiating. If ever, during differentiating, we have found any particular expression—in this example ax^{12}—we can find our way back to the y from which it was derived. The contrast between the two processes may be illustrated by the following illustration due to a well-known teacher. If a stranger were set down in Trafalgar Square, London, and told to find his way to Easton Station, he might find the task hopeless. But if he had previously been personally conducted from Easton Station to Trafalgar Square, it would be comparatively easy for him to find his way back to Easton Station.

Integration of the Sum or Difference of Two Functions

Let
$$\frac{dy}{dx} = x^2 + x^3$$

then
$$dy = x^2 dx + x^3 dx$$

There is no reason why we should not integrate each term separately; for, as may be seen in Chapter 6, we found that when we differentiated the sum of two separate functions, the derivative was simply the sum of the two separate differentiations. So, when we work backwards, integrating, the integration will be simply the sum of the two separate integrations.

Our instructions will then be:

$$\int dy = \int (x^2 + x^3)dx$$

$$= \int x^2 dx + \int x^3 dx$$

$$y = \tfrac{1}{3}x^3 + \tfrac{1}{4}x^4 + C$$

If either of the terms had been a negative quantity, the corresponding term in the integral would have also been negative. So that differences are as readily dealt with as sums.

How to Deal with Constant Terms

Suppose there is in the expression to be integrated a constant term—such as this:

$$\frac{dy}{dx} = x^n + b$$

This is laughably easy. For you have only to remember that when you differentiated the expression $y = ax$, the result was $\frac{dy}{dx} = a$. Hence, when you work the other way and integrate, the constant reappears multiplied by x. So we get

$$dy = x^n dx + b \cdot dx$$

$$\int dy = \int x^n dx + \int b\, dx$$

$$y = \frac{1}{n+1}x^{n+1} + bx + C$$

Here are a lot of examples on which to try your newly acquired powers.

Examples.

(1) Given $\dfrac{dy}{dx} = 24x^{11}$. Find y. *Ans.* $y = 2x^{12} + C$

(2) Find $\displaystyle\int (a + b)(x + 1)dx$. It is $(a + b)\displaystyle\int (x + 1)dx$

or $\quad (a + b)\left[\displaystyle\int x\, dx + \displaystyle\int dx\right]$ or $(a + b)\left(\dfrac{x^2}{2} + x\right) + C$

(3) Given $\dfrac{du}{dt} = gt^{\frac{1}{2}}$ Find u. *Ans.* $u = \tfrac{2}{3}gt^{\frac{3}{2}} + C$

(4) $\dfrac{dy}{dx} = x^3 - x^2 + x$ Find y.

$dy = (x^3 - x^2 + x)\, dx$

or $\quad dy = x^3 dx - x^2 dx + x\, dx;\ y = \displaystyle\int x^b dx - \displaystyle\int x^2 dx + \displaystyle\int x\, dx$

and $\quad y = \tfrac{1}{4}x^4 - \tfrac{1}{3}x^3 + \tfrac{1}{2}x^2 + C$

(5) Integrate $9.75x^{2.25}dx$. *Ans.* $y = 3x^{3.25} + C$

All these are easy enough. Let us try another case.

Let $\qquad\qquad\qquad \dfrac{dy}{dx} = ax^{-1}$

Proceeding as before, we will write

$$dy = ax^{-1} \cdot dx, \quad \int dy = a \int x^{-1}dx$$

Well, but what is the integral of $x^{-1}dx$?

If you look back amongst the results of differentiating x^2 and x^3 and x'', etc., you will find we never got x^{-1} from any one of them as the value of $\dfrac{dy}{dx}$. We got $3x^2$ from x^3; we got $2x$ from x^2;

we got 1 from x^1 that is, from x itself; but we did not get x^1 from x^0, for a very good reason. Its derivative (got by following the usual rule) is $0 \cdot 3 \, x^1$, and that multiplication by zero gives it zero value! Therefore when we now come to try to integrate $x^1 dx$, we see that it does not come in anywhere in the powers of x that are given by the rule:

$$\int x^n dx = \frac{1}{n+1} x^{n+1}$$

It is an exceptional case.

Well; but try again. Look through all the various derivatives obtained from various functions of x, and try to find amongst them x^{-1}. A sufficient search will show that we actually did get

$\dfrac{dy}{dx} = x^{-1}$ as the result of differentiating the function $y = \ln x$.

Then, of course, since we know that differentiating $\ln x$ gives us x^{-1}, we know that, by reversing the process, integrating $dy = x^{-1} dx$ will give us $y = \ln x$. But we must not forget the constant factor a that was given, nor must we omit to add the undetermined constant of integration. This then gives us as the solution to the present problem,

$$y = a \ln x + C$$

N.B.—Here note this very remarkable fact, that we could not have integrated in the above case if we had not happened to know the corresponding differentiation. If no one had found out that differentiating $\ln x$ gave x^{-1}, we should have been utterly stuck by the problem how to integrate $x^{-1} dx$. Indeed it should be frankly admitted that this is one of the curious features of the integral calculus:—that you can't integrate anything before the reverse process of differentiating something else has yielded that expression which you want to integrate.

Another Simple Case.

Find $\displaystyle\int (x + 1)(x + 2) dx$.

On looking at the function to be integrated, you remark that it is the product of two different functions of x. You could, you think, integrate $(x + 1)dx$ by itself, or $(x + 2)dx$ by itself. Of course you could. But what to do with a product? None of the differentiations you have learned have yielded you for the derivative a product like this. Failing such, the simplest thing is to multiply up the two functions, and then integrate. This gives us

$$\int (x^2 + 3x + 2)dx$$

And this is the same as

$$\int x^2 dx + \int 3x\,dx + \int 2\,dx$$

And performing the integrations, we get

$$\tfrac{1}{3}x^3 + \tfrac{3}{2}x^2 + 2x + C$$

Some Other Integral

Now that we know that integration is the reverse of differentiation, we may at once look up the derivatives we already know, and see from what functions they were derived. This gives us the following integrals ready made:

$$x^{-1}; \quad \int x^{-1}dx \quad = \ln |x| + C$$

$$\frac{1}{x+a}; \quad \int \frac{1}{x+a}dx \quad = \ln |x+a| + C$$

$$e^x; \quad \int e^x dx \quad = e^x + C$$

$$e^{-x}; \quad \int e^{-x}dx \quad = -e^{-x} + C$$

for if $y = \dfrac{-1}{e^x}$, $\dfrac{dy}{dx} = -\dfrac{e^x \times 0 - 1 \times e^x}{e^{2x}} = e^{-x}$

$$\sin x; \quad \int \sin x \, dx \quad = -\cos x + C$$

$$\cos x; \quad \int \cos x \, dx \quad = \sin x + C$$

Also we may deduce the following:

$$\ln x; \quad \int \ln x \, dx \quad = x(\ln x - 1) + C$$

for if $y = x \ln x - x$, $\dfrac{dy}{dx} = \dfrac{x}{x} + \ln x - 1 = \ln x$

$$\log_{10} x \int \log_{10} x \, dx = 0.4343x(\ln x - 1) + C$$

$$a^x; \quad \int a^x dx \quad = \dfrac{a^x}{\ln a} + C$$

$$\cos ax; \quad \int \cos ax \, dx \quad = \dfrac{1}{a} \sin ax + C$$

for if $y = \sin ax$, $\dfrac{dy}{dx} = a \cos ax$; hence to get $\cos ax$ one must differentiate $y = \dfrac{1}{a} \sin ax$

$$\sin ax; \quad \int \sin ax \, dx = -\dfrac{1}{a} \cos ax + C$$

Try also $\cos^2 \theta$; a little dodge will simplify matters:

$$\cos 2\theta = \cos^2 \theta - \sin^2 \theta = 2 \cos^2 \theta - 1$$

hence
$$\cos^2 \theta = \tfrac{1}{2}(\cos 2\theta + 1)$$

and
$$\int \cos^2 \theta \, d\theta = \tfrac{1}{2}\int (\cos 2\theta + 1)d\theta$$

$$= \tfrac{1}{2}\int \cos 2\theta \, d\theta + \tfrac{1}{2}\int d\theta$$

$$= \frac{\sin 2\theta}{4} + \frac{\theta}{2} + C.$$

See also the Table of Standard Forms at the back of the book. You should make such a table for yourself, putting in it only the general function which you have successfully differentiated and integrated. See to it that it grows steadily!

Exercises XVII

(See answers on page 306)

(1) Find $\int y \, dx$ when $y^2 = 4ax$.

(2) Find $\int \dfrac{3}{x^4} dx$.   (3) Find $\int \dfrac{1}{a} x^3 \, dx$.

(4) Find $\int (x^2 + a)dx$.   (5) Integrate $5x^{-\frac{7}{2}}$.

(6) Find $\int (4x^3 + 3x^2 + 2x + 1)dx$.

(7) If $\dfrac{dy}{dx} = \dfrac{ax}{2} + \dfrac{bx^2}{3} + \dfrac{cx^3}{4}$; find y.

(8) Find $\displaystyle\int\left(\frac{x^2+a}{x+a}\right)dx.$

(9) Find $\displaystyle\int(x+3)^3dx.$

(10) Find $\displaystyle\int(x+2)(x-a)dx.$

(11) Find $\displaystyle\int\left(\sqrt{x}+\sqrt[3]{x}\right)3a^2dx.$

(12) Find $\displaystyle\int\left(\sin\theta-\tfrac{1}{2}\right)\frac{d\theta}{3}.$

(13) Find $\displaystyle\int\cos^2 a\theta\, d\theta.$

(14) Find $\displaystyle\int\sin^2\theta\, d\theta.$

(15) Find $\displaystyle\int\sin^2 a\theta\, d\theta.$

(16) Find $\displaystyle\int e^{3x}\, dx.$

(17) Find $\displaystyle\int\frac{dx}{1+x}.$

(18) Find $\displaystyle\int\frac{dx}{1-x}.$

CHAPTER 19

On Finding Areas by Integrating

One use of the integral calculus is to enable us to ascertain the values of areas bounded by curves.

Let us try to get at the subject bit by bit.

Let AB be a curve, the equation to which is known. That is, y in this curve is some known function of x. (See Fig. 52.)

Think of a piece of the curve from the point P to the point Q.

Let a perpendicular PM be dropped from P, and another QN from the point Q. Then call $OM = x_1$ and $ON = x_2$, and the ordinates $PM = y_1$ and $QN = y_2$. We have thus marked out the area $PQNM$ that lies beneath the piece PQ. The problem is, *how can we calculate the value of this area?*

The secret of solving this problem is to conceive the area as being divided up into a lot of narrow strips, each of them being of the width dx. The smaller we take dx, the more of them there will be between x_1 and x_2. Now, the whole area is clearly equal to the sum of the areas of all such strips. Our business will then

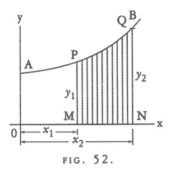

FIG. 52.

be to discover an expression for the area of any one narrow strip, and to integrate it so as to add together all the strips. Now think of any one of the strips. It will be like this: being bounded between two vertical sides, with a flat bottom dx, and with a slightly curved sloping top. Suppose we take its *average* height as being y; then, as its width is dx, its area will be ydx. And seeing that we may take the width as narrow as we please, if we only take it narrow enough its average height will be the same as the height at the middle of it. Now let us call the unknown value of the whole area S, meaning surface. The area of one strip will be simply a bit of the whole area, and may therefore be called dS. So we may write

$$\text{area of 1 strip} = dS = y \, dx$$

If then we add up all the strips, we get

$$\text{total area } S = \int dS = \int y \, dx$$

So then our finding S depends on whether we can integrate $y \, dx$ for the particular case, when we know what the value of y is as a function of x.

For instance, if you were told that for the particular curve in question $y = b + ax^2$, no doubt you could put that value into the expression and say: then I must find $\int (b + ax^2) dx$.

That is all very well; but a little thought will show you that something more must be done. Because the area we are trying to find is not the area under the whole length of the curve, but only the area limited on the left by *PM*, and on the right by *QN*, it follows that we must do something to define our area between those bounds.

This introduces us to a new notion, namely, that of *integrating between limits.* We suppose *x* to vary, and for the present purpose we do not require any value of *x* below x_1 (that is *OM*), nor any value of *x* above x_2 (that is *ON*). When an integral is to be thus defined between two limits, we call the lower of the two values *the inferior limit,* and the upper value *the superior limit.* Any integral so limited we designate as a *definite integral,* by way of distinguishing it from a *general integral* to which no limits are assigned.

In the symbols which give instructions to integrate, the limits are marked by putting them at the top and bottom respectively of the sign of integration. Thus the instruction

$$\int_{x=x_1}^{x=x_2} y \cdot dx$$

will be read: find the integral of $y \cdot dx$ between the inferior limit x_1 and the superior limit x_2.

Sometimes the thing is written more simply

$$\int_{x_1}^{x_2} y \cdot dx$$

Well, but *how* do you find an integral between limits when you have got these instructions?

Look again at Fig. 52. Suppose we could find the area under the larger piece of curve from *A* to *Q*, that is from $x = 0$ to $x = x_2$, naming the area *AQNO*. Then, suppose we could find the area under the smaller piece from *A* to *P*, that is from $x = 0$ to $x = x_1$, namely, the area *APMO*. If then we were to subtract the smaller area from the larger, we should have left as a remainder the area

PQNM, which is what we want. Here we have the clue as to what to do: the definite integral between the two limits is *the difference* between the integral worked out for the superior limit and integral worked out for the lower limit.

Let us then go ahead. First, find the integral thus:

$$\int y\,dx$$

and, as $y = b + ax^2$ is the equation to the curve (Fig. 52),

$$\int (b + ax^2)dx$$

is the integral which we must find

Doing the integration in question, we get

$$bx + \frac{a}{3}x^3 + C$$

and this will be the whole area from 0 up to any value of x that we may assign.

Therefore, the larger area up to the superior limit x_2 will be

$$bx_2 + \frac{a}{3}x_2{}^3$$

and the smaller area up to the inferior limit x_1 will be

$$bx_1 + \frac{a}{3}x_1{}^3$$

Now, subtract the smaller from the larger, and we get for the area S the value,

$$\text{area } S = b(x_2 - x_1) + \frac{a}{3}(x_2{}^3 - x_1{}^3)$$

This is the answer we wanted. Let us give some numerical values. Suppose $b = 10$, $a = 0.06$, and $x_2 = 8$ and $x_1 = 6$. Then the area S is equal to

$$10(8 - 6) + \frac{0.06}{3}(8^3 - 6^3)$$

$$= 20 + 0.02(512 - 216)$$

$$= 20 + 0.02 \times 296$$

$$= 20 + 5.92$$

$$= 25.92$$

Let us here put down a symbolic way of stating what we have ascertained about limits:

$$\int_{x=x_1}^{x=x_2} y\, dx = y_2 - y_1$$

where y_2 is the integrated value of $y\, dx$ corresponding to x_2, and y_1 that corresponding to x_1.

All integration between limits requires the difference between two values to be thus found. Also note that, in making the subtraction the added constant C has disappeared.

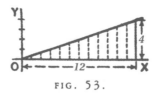

FIG. 53.

Examples.

(1) To familiarize ourselves with the process, let us take a case of which we know the answer beforehand. Let us find the area of the triangle (Fig. 53), which has base $x = 12$ and height $y = 4$. We know beforehand, from obvious mensuration, that the answer will come to 24.

Now, here we have as the "curve" a sloping line for which the equation is

212

$$y = \frac{x}{3}$$

The area in question will be

$$\int_{x=0}^{x=12} y \cdot dx = \int_{x=0}^{x=12} \frac{x}{3} \cdot dx$$

Integrating $\frac{x}{3} dx$, and putting down the value of the integral in square brackets with the limits marked above and below, we get

$$\text{area} = \left[\frac{1}{3} \cdot \frac{1}{2} x^2 + C \right]_{x=0}^{x=12}$$

$$= \left[\frac{x^2}{6} + C \right]_{x=0}^{x=12}$$

$$\text{area} = \left[\frac{12^2}{6} + C \right] - \left[\frac{0^2}{6} + C \right]$$

$$= \frac{144}{6} = 24$$

Note that, in dealing with definite integrals, the constant C always disappears by subtraction.

Let us satisfy ourselves about this rather surprising dodge of calculation, by testing it on a simple example, Get some squared paper, preferably some that is ruled in little squares of one-eighth or one-tenth inch each way. On this squared paper plot out the graph of this equation,

$$y = \frac{x}{3}$$

The values to be plotted will be:

x	0	3	6	9	12
y	0	1	2	3	4

213

The plot is given in Fig. 54.

Now reckon out the area beneath the curve *by counting the little squares* below the line, from $x = 0$ as far as $x = 12$ on the right. There are 18 whole squares and four triangles, each of which has an area equal to $1\frac{1}{2}$ squares; or, in total, 24 squares. Hence 24 is

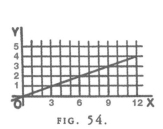

FIG. 54.

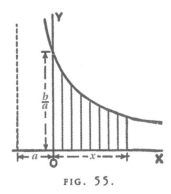

FIG. 55.

the numerical value of the integral of $\dfrac{x}{3}\ dx$ between the lower limit of $x = 0$ and the higher limit of $x = 12$.

As a further exercise, show that the value of the same integral between the limits of $x = 3$ and $x = 15$ is 36.

(2) Find the area, between limits $x = x_1$ and $x = 0$, of the curve $y = \dfrac{b}{x + a}$

$$\text{Area} = \int_{x=0}^{x=x_1} y \cdot dx = \int_{x=0}^{x=x_1} \frac{b}{x + a}\ dx$$

$$= b\left[\ln(x + a) + C\right]_0^{x_1}$$

$$= b[\ln(x_1 + a) + C - \ln(0 + a) - C]$$

$$= b \ln \frac{x_1 + a}{a}$$

Let it be noted that this process of subtracting one part from a larger to find the difference is really a common practice. How do

214

you find the area of a plane ring (Fig. 56), the outer radius of which is r_2 and the inner radius is r_1? You know from mensuration that the area of the outer circle is $\pi r_2{}^2$; then you find the area of the inner circle $\pi r_1{}^2$; then you subtract the latter from the former, and find area of ring $= \pi(r_2{}^2 - r_1{}^2)$; which may be written

$$\pi(r_2 + r_1)(r_2 - r_1)$$

= mean circumference of ring × width of ring.

(3) Here's another case—that of *the die-away curve*. Find the area between $x = 0$ and $x = a$ of the curve (Fig. 57) whose equation is

$$y = be^{-x}$$

$$\text{Area} = b \int_{x=0}^{x=a} e^{-x} \cdot dx$$

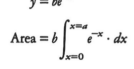

FIG. 56.

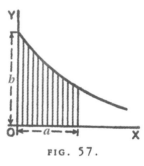

FIG. 57.

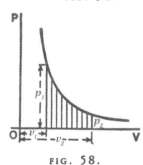

FIG. 58.

The integration gives

$$= b\left[-e^{-x}\right]_0^a$$

$$= b[-e^{-a} - (-e^{-0})]$$

$$= b(1 - e^{-a})$$

(4) Another example is afforded by the adiabatic curve of a perfect gas, the equation to which is $pv^n = c$, where p stands for pressure, v for volume, and n has the value 1.42 (Fig. 58).

Find the area under the curve (which is proportional to the work done in suddenly compressing the gas) from volume v_2 to volume v_1.

Here we have

$$\text{area} = \int_{v=v_1}^{v=v_2} cv^{-n} \cdot dv$$

$$= c\left[\frac{1}{1-n}\, v^{1-n}\right]_{v_1}^{v_2}$$

$$= c\,\frac{1}{1-n}(v_2^{\,1-n} - v_1^{\,1-n})$$

$$= \frac{-c}{0.42}\left(\frac{1}{v_2^{\,0.42}} - \frac{1}{v_1^{\,0.42}}\right)$$

An Exercise.

Prove the ordinary mensuration formula, that the area A of a circle whose radius is R, is equal to πR^2.

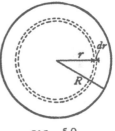

FIG. 59.

Consider an elementary zone or annulus of the surface (Fig. 59), of breadth dr, situated at a distance r from the center. We may consider the entire surface as consisting of such narrow zones, and the whole area A will simply be the integral of all such elementary zones from center to margin, that is, integrated from $r = 0$ to $r = R$.

We have therefore to find an expression for the elementary area dA of the narrow zone. Think of it as a strip of breadth dr, and of a length that is the periphery of the circle of radius r, that is, a length of $2\pi r$. Then we have, as the area of the narrow zone,

$$dA = 2\pi r\, dr$$

Hence the area of the whole circle will be:

$$A = \int dA = \int_{r=0}^{r=R} 2\pi r \cdot dr = 2\pi \int_{r=0}^{r=R} r \cdot dr$$

Now, the integral of $r \cdot dr$ is $-\dfrac{1}{2}r^2$. Therefore,

$$A = 2\pi\left[\tfrac{1}{2}r^2\right]_{r=0}^{r=R}$$

or $\qquad A = 2\pi\left[\tfrac{1}{2}R^2 - \tfrac{1}{2}(0)^2\right]$

whence $\quad A = \pi R^2$

Another Exercise.

Let us find the mean value of the positive part of the curve $y = x - x^2$, which is shown in Fig. 60. To find the mean ordinate, we shall have to find the area of the piece *OMN*, and then divide it by the length of the base *ON*. But before we can find the area we must ascertain the length of the base, so as to know up to what limit we are to integrate. At *N* the ordinate y has zero value; therefore, we must look at the equation and see what value of x will make $y = 0$. Now, clearly, if x is 0, y will also be 0, the curve passing through the origin *O*; but also, if $x = 1$, $y = 0$: so that $x = 1$ gives us the position of the point *N*.

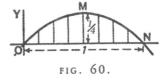

FIG. 60.

Then the area wanted is

$$= \int_{x=0}^{x=1}(x - x^2)dx = \left[\tfrac{1}{2}x^2 - \tfrac{1}{3}x^3\right]_0^1 = \left[\tfrac{1}{2} - \tfrac{1}{3}\right] - [0 - 0] = \tfrac{1}{6}$$

But the base length is 1.

Therefore, the average ordinate of the curve $= \tfrac{1}{6}$.

[*N.B.*—It will be a pretty and simple exercise in maxima and minima to find by differentiation what is the height of the maximum ordinate. It *must* be greater than the average.]

The mean ordinate of any curve, over a range from $x = 0$ to $x = x_1$, is given by the expression,

$$\text{mean } y = \frac{1}{x_1}\int_{x=0}^{x=x_1} y \cdot dx$$

217

If the mean ordinate be required over a distance not beginning at the origin but beginning at a point distant x_1 from the origin and ending at a point distant x_2 from the origin, the value will be

$$\text{mean } y = \frac{1}{x_2 - x_1} \int_{x=x_1}^{x=x_2} y \, dx$$

Areas in Polar Coordinates

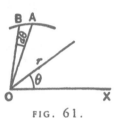

When the equation of the boundary of an area is given as a function of the distance r of a point of it from a fixed point O (see Fig. 61) called the *pole*, and of the angle which r makes with the positive horizontal direction OX, the process just explained can be applied just as easily, with a small modification. Instead of a strip of area, we consider a

FIG. 61.

small triangle OAB, the angle at O being $d\theta$, and we find the sum of all the little triangles making up the required area.

The area of such a small triangle is approximately $\dfrac{r \, d\theta}{2} \times r$;

hence the portion of the area included between the curve and two positions of r corresponding to the angles θ_1 and θ_2 is given by

$$\tfrac{1}{2} \int_{\theta=\theta_1}^{\theta=\theta_2} r^2 \, d\theta$$

Examples.

(1) Find the area of the sector of 1 radian in a circumference of radius a inch.

The polar equation of the circumference is evidently $r = a$. The area is

$$\tfrac{1}{2} \int_{\theta=0}^{\theta=1} a^2 \, d\theta = \frac{a^2}{2} \int_{\theta=0}^{\theta=1} d\theta = \frac{a^2}{2}$$

(2) Find the area of the first quadrant of the curve (known as a "cardioid"), the polar equation of which is

$$r = a(1 + \cos \theta)$$

218

$$\text{Area} = \tfrac{1}{2}\int_{\theta=0}^{\theta=\frac{\pi}{2}} a^2(1 + \cos\theta)^2 d\theta$$

$$= \frac{a^2}{2}\int_{\theta=0}^{\theta=\frac{\pi}{2}}(1 + 2\cos\theta + \cos^2\theta)d\theta$$

$$= \frac{a^2}{2}\left[\theta + 2\sin\theta + \frac{\theta}{2} + \frac{\sin 2\theta}{4}\right]_0^{\frac{\pi}{2}}$$

$$= \frac{a^2(3\pi + 8)}{8}$$

Volumes by Integration

What we have done with the area of a little strip of a surface, we can, of course, just as easily do with the volume of a little strip of a solid. We can add up all the little strips that make up the total solid, and find its volume, just as we have added up all the small little bits that made up an area to find the final area of the figure operated upon.

Examples.

(1) Find the volume of a sphere of radius r.

A thin spherical shell has for volume $4\pi x^2\,dx$ (see Fig. 59). Summing up all the concentric shells which make up the sphere, we have

$$\text{volume sphere} = \int_{x=0}^{x=r} 4\pi x^2\,dx = 4\pi\left[\frac{x^3}{3}\right]_0^r = \tfrac{4}{3}\pi r^3$$

We can also proceed as follows: a slice of the sphere, of thickness dx, has for volume $\pi y^2\,dx$ (see Fig. 62). Also x and y are related by the expression

$$y^2 = r^2 - x^2$$

Hence $\text{volume sphere} = 2\int_{x=0}^{x=r} \pi(r^2 - x^2)dx$

219

$$= 2\pi \left[\int_{x=0}^{x=r} r^2 \, dx - \int_{x=0}^{x=r} x^2 \, dx \right]$$

$$= 2\pi \left[r^2 x - \frac{x^3}{3} \right]_0^r = \frac{4\pi}{3} r^3$$

(2) Find the volume of the solid generated by the revolution of the curve $y^2 = 6x$ about the axis of x, between $x = 0$ and $x = 4$.

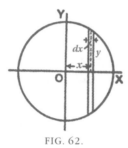

FIG. 62.

The volume of a slice of the solid is $\pi y^2 dx$.

Hence   volume $= \int_{x=0}^{x=4} \pi y^2 dx = 6\pi \int_{x=0}^{x=4} x \, dx$

$$= 6\pi \left[\frac{x^2}{2} \right]_0^4 = 48\pi = 150.8.$$

On Quadratic Means

In certain branches of physics, particularly in the study of alternating electric currents, it is necessary to be able to calculate the *quadratic mean* of a variable quantity. By "quadratic mean" is denoted the square root of the mean of the squares of all the values between the limits considered. Other names for the quadratic mean of any quantity are its "virtual" value, or its "R.M.S." (meaning root-mean-square) value. The French term is *valeur efficace*. If y is the function under consideration, and the quadratic mean is to be taken between the limits of $x = 0$ and $x = k$; then the quadratic mean is expressed as

$$\sqrt[2]{\frac{1}{k}\int_0^k y^2\,dx}$$

Examples.

(1) To find the quadratic mean of the function $y = ax$ (Fig. 63).

Here the integral is $\int_0^k a^2x^2\,dx$ which is $\frac{1}{3}a^2k^3$. Dividing by k and taking the square root, we have

$$\text{quadratic mean} = \frac{1}{\sqrt{3}}ak$$

Here the arithmetical mean is $\frac{1}{2}ak$; and the ratio of quadratic to arithmetical mean (this ratio is called the *form-factor*) is $2/\sqrt{3} = 2\sqrt{3}/3 = 1.1547.\ \ldots$

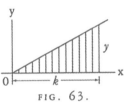

FIG. 63.

(2) To find the quadratic mean of the function $y = x^a$.

The integral is $\int_{x=0}^{x=k} x^{2a}\,dx$, that is $\dfrac{k^{2a+1}}{2a+1}$

Hence $\qquad \text{quadratic mean} = \sqrt{\dfrac{k^{2a}}{2a+1}}$

(3) To find the quadratic mean of the function $y = a^{\frac{x}{2}}$

The integral is $\int_{x=0}^{x=k} \left(a^{\frac{x}{2}}\right)^2 dx$, that is $\int_{x=0}^{x=k} a^x\,dx$

or $\qquad \left[\dfrac{a^x}{\ln a}\right]_{x=0}^{x=k}$, which is $\dfrac{a^k-1}{\ln a}$

Hence the quadratic mean is $\sqrt{\dfrac{a^k-1}{k \ln a}}$

221

Additional Problem Solving Examples

 If f and g are two continuous functions on the closed interval [a,b], then the area of the region bounded by the graphs of these two functions and the ordinates x = a and x = b is

$$A = \int_a^b [f(x)-g(x)]\,dx.$$

where $\qquad f(x) \geq 0 \quad \text{and} \quad f(x) \geq g(x)$

$$a \leq x \leq b$$

This formula applies whether the curves are above or below the x-axis.

The area below f(x) and above the x-axis is represented

by $\int_a^b f(x)$. The area between g(x) and the x-axis is represented

by $\int g(x)$.

Find the area of the region bounded by the curves $y = x^2$ and $y = \sqrt{x}$.

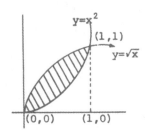

$$\text{Area} = A = \int_0^1 (\sqrt{x} - x^2)\,dx$$

$$= \int_0^1 \sqrt{x}\,dx - \int_0^1 x^2\,dx$$

$$= \left[\frac{2}{3} x^{\frac{3}{2}} - \frac{1}{3} x^3\right]_0^1$$

$$A = \left[\frac{2}{3} - \frac{1}{3}\right] = \frac{1}{3}$$

Q Find the area of the region bounded by the x-axis, the curve: $y = 6x - x^2$, and the vertical lines: x = 1 and x = 4.

A

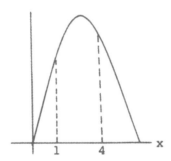

The limits of the integral which give the required area are x = 1 and x = 4. The function: $y = 6x - x^2$ is above the function y = 0 (the x-axis), therefore the area can be found by taking the integral of the upper function minus the lower function, or, $y = 6x - x^2$ minus y = 0, from x = 1 to x = 4. Therefore, we obtain:

$$A = \int_1^4 (6x - x^2) - 0\,dx = \left[3x^2 - \frac{x^3}{3}\right]_1^4$$

$$= \frac{80}{3} - \frac{8}{3} = 24.$$

Q Find the volume of the solid generated by revolving about the y-axis the region bounded by the parabola: $y^2 = 4x$, the y-axis and the line $y = 2$.

A

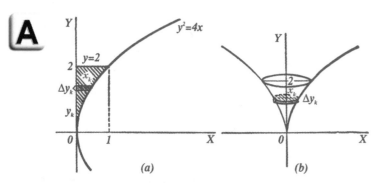

(a)   (b)

An element of volume is the disk generated by rotating a strip: x by dy, about the y-axis. The volume of the disk is: $dV = \pi x^2\, dy$.

Hence,

$$V = \pi \int_0^2 x^2\, dy$$

$$= \pi \int_0^2 \frac{1}{16} y^4\, dy = \frac{2}{5}\pi.$$

Shell Method

This method applies to cylindrical shells exemplified by

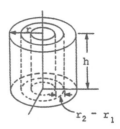

The volume of a cylindrical shell is

$$V = \pi r_2^2\, h - \pi r_1^2\, h$$

$$= \pi (r_2 + r_1)(r_2 - r_1) h$$

$$= 2\pi \left(\frac{r_2 + r_1}{2} \right)(r_2 - r_1) h$$

where r_1 = inner radius

           r_2 = outer radius

           h = height.

Let $r = \frac{r_1 + r_2}{2}$ and $\Delta r = r_2 - r_1$, then

the volume of a shell becomes

$$\boxed{V = 2\pi r h \Delta r}$$

The thickness of the shell is represented by Δr and the average radius of the shell by r.

Thus,

$$\boxed{V = 2\pi \int_a^b x f(x)\, dx}$$

is the volume of a solid generated by revolving a region about the y-axis. This is illustrated below:

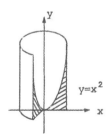

$y = x^2$

Q Find the volume of the solid generated by revolving about the y-axis the region bounded by the parabola: $y = -x^2 + 6x - 8$, and the x-axis.

A

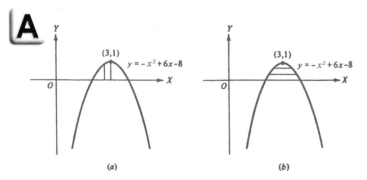

(a)   (b)

Method 1. We use the method of cylindrical shells. The curve:
$$y = -x^2 + 6x - 8,$$
cuts the x-axis at $x = 2$ and $x = 4$.

The cylindrical shells are generated by the strip formed by the two lines parallel to the y-axis, at distances x and $x + \Delta x$ from the y-axis, $2 \le x \le 4$, as shown in figure (a). When this strip is revolved about the y-axis, it generates a cylindrical shell of average height y*, $y \le y^* \le y + \Delta y$, thickness Δx, and average radius x*, $x < x^* \le x + \Delta x$. The voume of this element is:
$$\Delta V = 2\pi x^* \, y^* \, \Delta x,$$
where $2\pi x^* \, y^*$ is the surface area. Expressing y in terms of x and passing to the limits, the sum of the volumes of all such cylindrical shells is the integral:

$$V = 2\pi \int_2^4 x(-x^2 + 6x - 8) \, dx$$

$$= 2\pi \int_2^4 (-x^3 + 6x^2 - 8x) \, dx$$

$$= 2\pi \left(-\frac{x^4}{4} + 2x^3 - 4x^2 \right) \Big|_2^4$$

$$= 2\pi \Big((-64 + 128 - 64) - (-4 + 16 - 16) \Big)$$

$$= 8\pi.$$

Method 2. This can also be thought of as the volume comprising a series of concentric washers with variable outer and inner radii, as sectionally shown in figure (b). The variable radii are as follows: Since

$$y = -x^2 + 6x - 8,$$

we solve for x.

To complete the square, we require a 9, so that

$$x^2 - 6x + 9$$

constitutes a perfect square. Rewriting the equation,

$$x^2 - 6x + 9 - 9 + 8 = -y.$$

$$x^2 - 6x + 9 = 1 - y.$$

$$(x - 3)^2 = 1 - y.$$

Therefore,

$$x = 3 \pm \sqrt{1-y}$$

which shows the washers, y units from the x-axis, have an

inner radius: $x_{in} = 3 - \sqrt{1-y}$, and an

outer radius: $x_o = 3 + \sqrt{1-y}$.

(The particular one on the x-axis has $x_{in} = 2$ and $x_o = 4$.)

The volume of this washer with thickness dy is:

$$dV = \pi (x_o^2 - x_{in}^2)\, dy,$$

or $\quad dV = \pi \left[(x_o + x_{in})(x_o - x_{in}) \right] dy.$

Substituting the values for x_o and x_{in},

$$dV = \pi \bigg(\big((3+\sqrt{1-y})+(3-\sqrt{1-y})\big).$$

$$\big((3+\sqrt{1-y})-(3-\sqrt{1-y})\big)\bigg) dy$$

$$= \pi(12 \sqrt{1-y}) \, dy.$$

Since y varies from 0 to 1, the desired volume is:

$$V = 12\pi \int_0^1 (1-y)^{\frac{1}{2}} \, dy$$

$$= 12\pi \left(-\frac{2}{3}(1-y)^{\frac{3}{2}} \,\bigg|_0^1 \right. = 8\pi.$$

Parallel Cross Sections

A cross section of a solid is a region formed by the intersection of a solid by a plane. This is illustrated below:

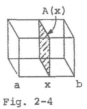

Fig. 2-4

228

If x is a continuous function on the interval [a,b], then the volume of the cross sectional area A(x) is

$$V = \int_a^b A(x)dx.$$

Rotate the curve y = 2x about the line x = 4 and find the volume produced by the rotation of the shaded portion.

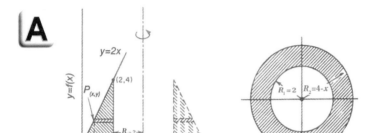

Area of washer $= \pi\left(R_2{}^2 - R_1{}^2\right)$

Increment of volume, $dV = \pi\left(R_2{}^2 - R_1{}^2\right)$ dy.

$V = \int_a^b \pi\left(R_2{}^2 - R_1{}^2\right)$ dy, where a and b are limits of y.

$V = \int_0^4 \pi\left((4-x)^2 - 2^2\right)$ dy

$\quad = \int_0^4 \pi\ (16 - 8x + x^2 - 4)$ dy

$\quad = \pi \int_0^4 (12 - 8x + x^2)$ dy.

But the x terms must be in terms of y because of dy. We have:

$x = \dfrac{y}{2}$, from the equation of the curve. Substituting,

$$V = \pi \int_0^4 \left(12 - 4y + \dfrac{y^2}{4} \right) dy$$

$$= \pi \left(12y - \dfrac{4y^2}{2} + \dfrac{y^3}{12} \right)_0^4$$

$$= \pi \left(48 - 32 + - \dfrac{16}{3} \right) = \dfrac{64\pi}{3}.$$

This is the washer method. The important point to remember is to obtain the radii from the center line of rotation.

Exercises XVIII

(See answers on page 306)

(1) Find the area of the curve $y = x^2 + x + 5$ between $x = 0$ and $x = 6$, and the mean ordinate between these limits.

(2) Find the area of the parabola $y = 2a\sqrt{x}$ between $x = 0$ and $x = a$. Show that it is two-thirds of the rectangle of the limiting ordinate and of its abscissa.

(3) Find the area of the portion of a sine curve between $x = 0$ and $x = \pi$, and the mean ordinate.

(4) Find the area of the portion of the curve $y = \sin^2 x$ from $0°$ to $180°$, and find the mean ordinate.

(5) Find the area included between the two branches of the curve $y = x^2 \pm x^{\frac{5}{2}}$ from $x = 0$ to $x = 1$, also the area of the positive portion of the lower branch of the curve (Fig. 30).

(6) Find the volume of a cone of radius of base r, and of height h.

(7) Find the area of the curve $y = x^3 - \ln x$ between $x = 0$ and $x = 1$.

230

(8) Find the volume generated by the curve $y = \sqrt{1 + x^2}$, as it revolves about the axis of x, between $x = 0$ and $x = 4$.

(9) Find the volume generated by a sine curve between $x = 0$ and $x = \pi$, revolving about the axis of x.

(10) Find the area of the portion of the curve $xy = a$ included between $x = 1$ and $x = a$. Find the mean ordinate between these limits.

(11) Show that the quadratic mean of the function $y = \sin x$, between the limits of 0 and π radians, is $\dfrac{\sqrt{2}}{2}$. Find also the arithmetical mean of the same function between the same limits; and show that the form-factor is $=1.11$.

(12) Find the arithmetical and quadratic means of the function $x^2 + 3x + 2$, from $x = 0$ to $x = 3$.

(13) Find the quadratic mean and the arithmetical mean of the function $y = A_1 \sin x + A_3 \sin 3x$ between $x = 0$ and $x = 2\pi$.

(14) A certain curve has the equation $y = 3.42 e^{0.21x}$. Find the area included between the curve and the axis of x, from the ordinate at $x = 2$ to the ordinate at $x = 8$. Find also the height of the mean ordinate of the curve between these points.

(15) The curve whose polar equation is $r = a(1 - \cos \theta)$ is known as the cardioid. Show that the area enclosed by the axis and the curve between $\theta = 0$ and $\theta = 2\pi$ radians is equal to 1.5 times that of the circle whose radius is a.

(16) Find the volume generated by the curve

$$y = \pm \frac{x}{6} \sqrt{x(10 - x)}$$

rotating about the axis of x.

CHAPTER 20

Dodges, Pitfalls, and Triumphs

Why We Study this Topic

Complex functions can be difficult to integrate, as noted in previous chapters. This chapter describes various techniques through which integration can be simplified. It takes practice to know which technique to use, and how to apply it effectively. Practice is often what one must do to learn math.

Dodges. A great part of the labor of integrating things consists in licking them into some shape that can be integrated. The books—and by this is meant the serious books—on the integral calculus are full of plans and methods and dodges and artifices for this kind of work. The following are a few of them.

Integration by Parts. This name is given to a dodge, the formula for which is

$$\int u\ dx = ux - \int x\ du + C$$

It is useful in some cases that you can't tackle directly, for it shows that if in any case $\int x\ du$ can be found, then $\int u\ dx$ can also

232

be found. The formula can be deduced as follows.

$$d(ux) = u\, dx + x\, du$$

which may be written

$$u\, dx = d(ux) - x\, du$$

which by direct integration gives the above expression.

Examples.

(1) Find $\displaystyle\int w \cdot \sin w\, dw$

Write $u = w$, and dx for $\sin w \cdot dw$. We shall then have $du = dw$,

while $x = \displaystyle\int \sin w \cdot dw = -\cos w.$

Putting these into the formula, we get

$$\int w \cdot \sin w\, dw = w(-\cos w) - \int -\cos w\, dw$$

$$= -w \cos w + \sin w + C.$$

(2) Find $\displaystyle\int xe^x dx.$

Write $\qquad\qquad u = x \quad dv = e^x dx$

then $\qquad\qquad du = dx \quad v = e^x$

and $\qquad\displaystyle\int xe^x dx = xe^x - \int e^x dx$ (by the formula)

$$= xe^x - e^x + C = e^x(x - 1) + C$$

(3) Try $\displaystyle\int \cos^2\theta\, d\theta.$

$$u = \cos\theta \qquad\qquad dx = \cos\theta\, d\theta$$

Hence $\qquad du = -\sin\theta\, d\theta \qquad x = \sin\theta$

233

$$\int \cos^2\theta \; d\theta = \cos\theta \sin\theta + \int \sin^2\theta \; d\theta$$

$$= \frac{2\cos\theta\sin\theta}{2} + \int (1 - \cos^2\theta)d\theta$$

$$= \frac{\sin 2\theta}{2} + \int d\theta - \int \cos^2\theta \; d\theta$$

Hence
$$2\int \cos^2\theta \; d\theta = \frac{\sin 2\theta}{2} + \theta + 2C$$

and
$$\int \cos^2\theta \; d\theta = \frac{\sin 2\theta}{4} + \frac{\theta}{2} + C$$

(4) Find $\int x^2 \sin x \; dx$.

Write $\qquad\qquad u = x^2 \qquad dv = \sin x \; dx$

then $\qquad\qquad du = 2x \; dx \qquad v = -\cos x$

$$\int x^2 \sin x \; dx = -x^2 \cos x + 2\int x \cos x \; dx$$

Now find $\int x \cos x \; dx$, integrating by parts (as in Example 1 above):

$$\int x \cos x \; dx = x \sin x + \cos x + C$$

Hence $\int x^2 \sin x \; dx = -x^2 \cos x + 2x \sin x + 2 \cos x + C'$

$$= (2 - x^2) \cos x + 2x \sin x + C'$$

(5) Find $\int \sqrt{1-x^2}\,dx$.

Write $\qquad u = \sqrt{1-x^2}, \quad dx = dv$

then $\qquad du = -\dfrac{x\,dx}{\sqrt{1-x^2}}$ (see Chap. IX)

and $x = v$; so that

$$\int \sqrt{1-x^2}\,dx = x\,\sqrt{1-x^2} + \int \frac{x^2\,dx}{\sqrt{1-x^2}}$$

Here we may use a little dodge, for we can write

$$\int \sqrt{1-x^2}\,dx = \int \frac{(1-x^2)\,dx}{\sqrt{1-x^2}} = \int \frac{dx}{\sqrt{1-x^2}} - \int \frac{x^2\,dx}{\sqrt{1-x^2}}$$

Adding these two last equations, we get rid of $\int \dfrac{x^2\,dx}{\sqrt{1-x^2}}$,
and we have

$$2\int \sqrt{1-x^2}\,dx = x\sqrt{1-x^2} + \int \frac{dx}{\sqrt{1-x^2}}$$

Do you remember meeting $\dfrac{dx}{\sqrt{1-x^2}}$? It is got by differentiating $y = \text{arc sin } x$; hence its integral is arc sin x, and so

$$\int \sqrt{1-x^2}\,dx = \frac{x\sqrt{1-x^2}}{2} + \tfrac{1}{2}\,\text{arc sin } x + C$$

You can try now some exercises by yourself; you will find some
at the end of this chapter.

Substitution. This is the same dodge as explained in Chapter 9.
Let us illustrate its application to integration by a few examples.

(1) $\int \sqrt{3+x}\, dx$

Let $\qquad\qquad\qquad u = 3+x, \quad du = dx$

replace: $\qquad \int u^{\frac{1}{2}}\, du = \frac{2}{3} u^{\frac{3}{2}} + C = \frac{2}{3}(3+x)^{\frac{3}{2}} + C$

(2) $\int \dfrac{dx}{e^x + e^{-x}}$

Let $\qquad\qquad u = e^x, \dfrac{du}{dx} = e^x, \text{ and } dx = \dfrac{du}{e^x}$

so that $\quad \displaystyle\int \dfrac{dx}{e^x + e^{-x}} = \int \dfrac{du}{e^x(e^x + e^{-x})} = \int \dfrac{du}{u\left(u + \dfrac{1}{u}\right)} = \int \dfrac{du}{u^2 + 1}$

$\dfrac{du}{1+u^2}$ is the result of differentiating arc tan u.

Hence the integral is arc tan $e^x + C$.

(3) $\int \dfrac{dx}{x^2 + 2x + 3} = \int \dfrac{dx}{x^2 + 2x + 1 + 2} = \int \dfrac{dx}{(x+1)^2 + (\sqrt{2})^2}.$

Let $\qquad\qquad u = x+1, \quad du = dx;$

then the integral becomes $\displaystyle\int \dfrac{du}{u^2 + (\sqrt{2})^2};$ but $\dfrac{du}{u^2 + a^2}$ is the re-

sult of differentiating $\dfrac{1}{a}$ arc tan $\dfrac{u}{a}$.

Hence one has finally $\dfrac{1}{\sqrt{2}}$ arc tan $\dfrac{x+1}{\sqrt{2}} + C$ for the value of the given integral.

Rationalization, and *Factorization of Denominator* are dodges applicable in special cases, but they do not admit of any short or general explanation. Much practice is needed to become familiar with these preparatory processes.

The following example shows how the process of splitting into partial fractions, which we learned in Chapter 13, can be made use of in integration.

Take $\int \dfrac{dx}{x^2 + 2x - 3}$; if we split $\dfrac{1}{x^2 + 2x - 3}$ into partial fractions, this becomes:

$$\frac{1}{4}\left[\int \frac{dx}{x-1} - \int \frac{dx}{x+3}\right] = \frac{1}{4}\,[\ln(x-1) - \ln(x+3)] + C$$

$$= \frac{1}{4}\ln\frac{x-1}{x+3} + C$$

Notice that the same integral can be expressed sometimes in more than one way (which are equivalent to one another).

Pitfalls. A beginner is liable to overlook certain points that a practised hand would avoid; such as the use of factors that are equivalent to either zero or infinity, and the occurence of indeterminate quantities such as $\frac{0}{0}$. There is no golden rule that will meet every possible case. Nothing but practice and intelligent care will avail. An example of a pitfall which had to be circumvented arose in Chapter 18, when we came to the problem of integrating $x^{-1}\,dx$.

Triumphs. By triumphs must be understood the successes with which the calculus has been applied to the solution of problems otherwise intractable. Often in the consideration of physical relations one is able to build up an expression for the law governing the interaction of the parts or of the forces that govern them, such expression being naturally in the form of a *differential equation*, that is an equation containing derivatives with or without other algebraic quantities. And when such a differential equation has been found, one can get no further until it has been integrated. Generally it is much easier to state the appropriate differential equation than to solve it: the real trouble begins then only when one wants to integrate, unless indeed the equation is seen to possess some standard form of which the integral is known, and then the triumph is easy. The equation which results from integrating a differential equation is called* its "solution"; and it is quite astonishing how in many cases the solution looks as if it had no relation to the differential equation of which it is the integrated form. **The solution often seems as different from the original expression as a butterfly does from a caterpillar that it was.** Who would have supposed that such an innocent thing as

$$\frac{dy}{dx} = \frac{1}{a^2 - x^2}$$

could blossom out into

$$y = \frac{1}{2a} \ln \frac{a+x}{a-x} + C?$$

yet the latter is the *solution* of the former.

As a last example, let us work out the above together.

By partial fractions,

$$\frac{1}{a^2 - x^2} = \frac{1}{2a(a+x)} + \frac{1}{2a\,(a-x)}$$

$$dy = \frac{dx}{2a\,(a+x)} + \frac{dx}{2a\,(a-x)}$$

$$y = \frac{1}{2a} \left(\int \frac{dx}{a+x} + \int \frac{dx}{a-x} \right)$$

$$= \frac{1}{2a} [\ln (a+x) - \ln (a-x)] + C$$

$$= \frac{1}{2a} \ln \frac{a+x}{a-x} + C$$

---

*This means that the actual result of solving it is called its "solution". But many mathematicians would say, with Professor A.R. Forsyth, "every differential equation *is considered as solved* when the value of the dependent variable is expressed as a function of the independent variable by means either of known functions, or of integrals, whether the integrations in the latter can or cannot be expressed in terms of functions already known."

Not a very difficult metamorphosis!

There are whole treatises, such as George Boole's *Differential Equations*, devoted to the subject of finding the "solutions" for different original forms.

Additional Problem Solving Examples

Q Evaluate the expression: $\int_0^3 x\sqrt{1+x}\ dx$.

A We wish to convert the given integral into a form, to which we can apply the formula for $\int u^n du$. To evaluate the indefinite integral $\int x\sqrt{1+x}\ dx$, we let

$$u = \sqrt{1+x}, u^2 = 1 + x, x = u^2 - 1, dx = 2u\ du.$$

Substituting, we have:

$$\int x\sqrt{1+x}\ dx = \int (u^2 - 1)\ u\ (2u\ du)$$
$$= 2\int (u^4 - u^2)\ du.$$

We can now apply the formula for $\int u^n du$, and we obtain:

$$\frac{2}{5}u^5 - \frac{2}{3}u^3 + C = \frac{2}{5}(1+x)^{\frac{5}{2}} - \frac{2}{3}(1+x)^{\frac{3}{2}} + C,$$

by substitution. Therefore, the definite integral

$$\int_0^3 x\sqrt{1+x}\ dx = \frac{2}{5}(1+x)^{\frac{5}{2}} - \frac{2}{3}(1+x)^{\frac{3}{2}}\Big]_0^3$$
$$= \frac{2}{5}(4)^{\frac{5}{2}} - \frac{2}{3}(4)^{\frac{3}{2}} - \frac{2}{5}(1)^{\frac{5}{2}} + \frac{2}{3}(1)^{\frac{3}{2}}$$
$$= \frac{64}{5} - \frac{16}{3} - \frac{2}{5} + \frac{2}{3}$$
$$= \frac{116}{15}.$$

Change of Variables

Q Evaluate $\displaystyle\int_0^1 x(1+x)^{\frac{1}{2}}\,dx$

Let $u = 1+x$, $du = dx$, $x = u-1$

A $\displaystyle\int_0^1 x(1+x)^{\frac{1}{2}} = \int_1^2 (u-1)u^{\frac{1}{2}}\,du$.

*Notice the change in the limits for x=0, u=1 and for x=1 u=2.

$$\int_1^2 (u-1)u^{\frac{1}{2}}\,du = \int_1^2 u^{3/2} - u^{\frac{1}{2}}\,du$$

$$= 2/5\,u^{5/2} - 2/3\,u^{3/2}\Big|_1^2$$

$$= [(2/5)\sqrt{32} - (2/3)\sqrt{8})] - \left(\frac{2}{5} - \frac{2}{3}\right)$$

$$= \frac{4\sqrt{2}}{15} - \frac{4}{15} = \frac{4}{15}(\sqrt{2} - 1).$$

Q Evaluate the expression: $\displaystyle\int_1^2 \frac{x}{(1+2x)^3}\,dx$.

A This integral is difficult because of the expression: $1 + 2x$, in the denominator. Hence we choose our substitution to eliminate this expression. We let

$$u = 1 + 2x, \text{ then } x = \frac{u-1}{2} \text{ and } dx = \frac{1}{2}\,du.$$

Now

$u = 3$ when $x = 1$

$u = 5$ when $x = 2$,

giving us new limits. Using he substitution, we obtain:

$$\int_1^2 \frac{x}{(1+2x)^3}\,dx = \int_3^5 \left(\frac{\frac{u-1}{2}}{u^3}\right)\left(\frac{1}{2}\right)du$$

$$= \frac{1}{4}\int_3^5 \left(\frac{1}{u^2} - \frac{1}{u^3}\right)du.$$

We can now use the formula for $\int u^n du$ on both terms of the integrand, obtaining:

$$\frac{1}{4}\left[-\frac{1}{u} + \frac{1}{2u^2}\right]_3^5 = \frac{11}{450}.$$

Integration of Parts

Summary: This method is based on the formula

$$d(uv) = u\,dv + v\,du.$$

The corresponding integration formula,

$$uv = \int u\,dv + \int v\,du, \text{ is applied in the form}$$

$$\boxed{\int u\,dv = uv - \int v\,du}$$

This procedure involves the identification of u and dv and their manipulation into the form of the latter equation. v must be easily determined. If a definite integral is

$$\int_a^b u\frac{dv}{dx}\,dx = uv\Big]_a^b - \int_a^b v\frac{du}{dx}\,dx.$$

Evaluate $\int x\cos x\,dx$

$u = x \quad dv = \cos x\,dx$

$du = dx \quad v = \sin x$

$$\int x\cos x\,dx = x\sin x - \int \sin x\,dx$$

241

$$= x \sin x - (-\cos x) + c$$
$$= x \sin x + \cos x + c$$

 Integrate by parts the expression: $\dfrac{dy}{dx} = x^2 \ln x.$

$$dy = x^2 \ln x \cdot dx.$$
$$y = \int x^2 \ln x \, dx.$$

To integrate by parts we use the equation:

$\int udv = uv - \int vdu$.

Now, let $u = \ln \cdot x$.

Then, $du = \dfrac{1}{x} \cdot dx.$

Let $dv = x^2 \cdot dx.$

Then, $v = \int dv = \int x^2 \, dx = \dfrac{x^3}{3}$,

by use of the formula for $\int u^n du$. Substituting into the above equation, we have:

$$\overset{u}{y = \int \ln x} \cdot \overset{dv}{x^2 \cdot dx} = \dfrac{x^3}{3} \ \ln x - \int \dfrac{x^3}{3} \cdot \dfrac{1}{x} \cdot dx.$$

We can now integrate

$$\int \dfrac{x^3}{3} \cdot \dfrac{1}{x} \ \cdot dx,$$

by using the formula for $\int u^n du$, with $u = x$, $du = dx$, and $n = 2$. Doing this, we obtain:

$$y = \dfrac{x^3}{3} \ln x - \dfrac{x^3}{9} + C.$$

242

Trigonometric Integrals

Summary: Integrals of the form $\int \sin^n x \; dx$ or $\int \cos^n x \, dx$ can be evaluated without resorting to integration by parts. This is done in the following manner:

We write $\int \sin^n x \; dx = \int \sin^{n-1} \sin x \; dx$, **if n is odd.**

Since the integer n-1 is even, we may then use the fact that $\sin^2 x = 1-\cos^2 x$ to obtain a form which is easier to integrate.

Example: $\int \sin^5 x \; dx = \int \sin^4 x \; \sin x \; dx$

$$= \int (\sin^2 x)^2 \; \sin x \; dx$$

but $\sin^2 x = 1 - \cos^2 x$.

Hence, $\int \sin^5 x \; dx = \int (1-\cos^2 x)^2 \sin x \; dx$

$$= \int (1-2\cos^2 x + \cos^4 x)\sin x \; dx$$

Substitute $u = \cos x$, $du = -\sin x \; dx$

$$= -\int(1-2u^2+u^4)du = -u + \frac{2}{3}u^3 - \frac{u^5}{5}$$

$$= -\cos x + \frac{2}{3}\cos^3 x - \frac{1}{5}\cos^5 x + c.$$

A similar technique can be employed for odd powers of cos x.

If the integrand is $\sin^n x$ or $\cos^n x$ and n is even, then the half angle formulas,

$$\sin^2 x = \frac{1-\cos 2x}{2} \text{ or }$$
$$\cos^2 x = \frac{1+\cos 2x}{2}$$

243

may be used to simplify the integrand.

Example: $\int \cos^2 x \; dx = \frac{1}{2} \int (1 + \cos 2x) dx$

$$= \frac{1}{2} x + \frac{1}{4} \sin 2x + c$$

Integrate the expression: $\int \cos^2 x \sin x \; dx$.

In evaluating this integral we use the formula:

$$\int u^n \, du = \frac{u^{n+1}}{n+1} \; .$$

Let $u = \cos x$, $du = -\sin x \; dx$, and $n = 2$. Applying the formula, we obtain:

$$\int \cos^2 x \sin x \; dx = -\frac{\cos^3 x}{3} + C.$$

Exercises XIX

(See answers on page 307)

(1) Find $\int \sqrt{a^2 - x^2} \, dx$.

(2) Find $\int x \ln x \; dx$.

(3) Find $\int x^a \ln x \; dx$.

(4) Find $\int e^x \cos e^x \; dx$.

(5) Find $\int \frac{1}{x} \cos (\ln x) \; dx$.

(6) Find $\int x^2 e^x \; dx$.

(7) Find $\int \frac{(\ln x)^a}{x} \; dx$.

(8) Find $\int \frac{dx}{x \ln x}$.

(9) Find $\int \frac{5x + 1}{x^2 + x - 2} dx$.

(10) Find $\int \frac{(x^2 - 3) dx}{x^3 - 7x + 6}$.

(11) Find $\int \frac{b \; dx}{x^2 - a^2}$.

(12) Find $\int \frac{4x \; dx}{x^4 - 1}$.

(13) Find $\displaystyle\int \frac{dx}{1-x^4}$.

(14) Find $\displaystyle\int \frac{x\,dx}{\sqrt{a^2-b^2x^2}}$.

(15) Use the substitution $\dfrac{1}{x} = \dfrac{b}{a}\cosh u$ to show that

$$\int \frac{dx}{x\sqrt{a^2-b^2x^2}} = \frac{1}{a}\ln\frac{a-\sqrt{a^2-b^2x^2}}{x} + C$$

CHAPTER 21

Finding Solutions To Differential Equations

In this chapter we go to work finding solutions to some important differential equations, using for this purpose the processes shown in the preceding chapters.

The beginner, who now knows how easy most of those processes are in themselves, will here begin to realize that integration is *an art*. As in all arts, so in this, facility can be acquired only by diligent and regular practice. Those who would attain that facility must work out examples, and more examples, and yet more examples, such as are found abundantly in all the regular treatises on the calculus. Our purpose here must be to afford the briefest introduction to serious work.

Example 1. Find the solution of the differential equation

$$ay + b\frac{dy}{dx} = 0$$

Transposing, we have

$$b\frac{dy}{dx} = -ay$$

Now the mere inspection of this relation tells us that we have got to do with a case in which $\frac{dy}{dx}$ is proportional to y. If we think of the curve which will represent y as a function of x, it will be such that its slope at any point will be proportional to the ordinate at that point, and will be a negative slope if y is positive. So obviously the curve will be a die-away curve, and the solution will contain e^{-x} as a factor. But, without presuming on this bit of sagacity, let us go to work.

As both y and dy occur in the equation and on opposite sides, we can do nothing until we get both y and dy to one side, and dx to the other. To do this, we must split our usually inseparable companions dy and dx from one another.

$$\frac{dy}{y} = -\frac{a}{b}\,dx$$

Having done the deed, we now can see that both sides have got into a shape that is integrable, because we recognize $\dfrac{dy}{y}$, or $\dfrac{1}{y}\,dy$,

as a differential that we have met with when differentiating logarithms. So we may at once write down the instructions to integrate,

$$\int\frac{dy}{y} = \int -\frac{a}{b}dx$$

and doing the two integrations, we have:

$$\ln y = -\frac{a}{b}x + \ln C$$

where $\ln C$ is the yet undetermined constant* of integration. Then, delogarizing, we get:

$$y = Ce^{-\frac{a}{b}x}$$

which is *the solution* required. Now, this solution looks quite unlike the original differential equation from which it was constructed: yet to an expert mathematician they both convey the same information as to the way in which y depends on x.

---

*We may write down any form of constant as the "constant of integration", and the form $\ln C$ is adopted here by preference, because the other terms in this line of equation are, or are treated as logarithms; and it saves complications afterward if the added constant be *of the same kind*.

Now, as to the C, its meaning depends on the initial value of y. For if we put $x = 0$ in order to see what value y then has, we find that this makes $y = Ce^{-0}$; and as $e^{-0} = 1$, we see that C is nothing else than the particular value* of y at starting. This we may call y_0 and so write the solution as

$$y = y_0 e^{-\frac{a}{b}x}$$

Example 2.
Let us take as an example to solve

$$ay + b\frac{dy}{dx} = g$$

where g is a constant. Again, inspecting the equation will suggest, (1) that somehow or other e^x will come into the solution, and (2) that if at any part of the curve y becomes either a maximum or a minimum, so that $\frac{dy}{dx} = 0$, then y will have the value $= \frac{g}{a}$ But let us go to work as before, separating the differentials and trying to transform the thing into some integrable shape.

$$b\frac{dy}{dx} = g - ay$$

$$\frac{dy}{dx} = \frac{a}{b}\left(\frac{g}{a} - y\right)$$

$$\frac{dy}{y - \frac{g}{a}} = -\frac{a}{b}dx.$$

Now we have done our best to get nothing but y and dy on one side, and nothing but dx on the other. But is the result on the left side integrable?

---

*Compare what was said about the "constant of integration", with reference to Fig. 48, and Fig. 51.

It is of the same form as the result in Chapter 14, so, writing the instructions to integrate, we have:

$$\int \frac{dy}{y - \frac{g}{a}} = -\int \frac{a}{b} dx$$

and, doing the integration, and adding the appropriate constant,

$$\ln \left(y - \frac{g}{a} \right) = -\frac{a}{b} x + \ln C$$

whence

$$y - \frac{g}{a} = Ce^{-\frac{a}{b}x}$$

and finally,

$$y = \frac{g}{a} + Ce^{-\frac{a}{b}x}$$

which is *the solution.*

If the condition is laid down that $y = 0$ when $x = 0$ we can find C; for then the exponential becomes $=1$; and we have

$$0 = \frac{g}{a} + C$$

or

$$C = -\frac{g}{a}$$

Putting in this value, the solution becomes

$$y = \frac{g}{a}(1 - e^{-\frac{a}{b}x})$$

But further, if x grows infinitely, y will grow to a maximum; for when $x = \infty$, the exponential $= 0$, giving $y_{max.} = \frac{g}{a}$. Substituting this, we get finally

$$y = y_{max.}(1 - e^{-\frac{a}{b}x})$$

This result is also of importance in physical science.

Before proceeding to the next example, it is necessary to discuss two integrals which are of great importance in physics and engineering. These seem to be very elusive as, when either of them is tackled, it turns partly into the other. Yet this very fact helps us to determine their values. Let us denote these integrals by S and C, where

$$S = \int e^{pt} \sin kt \, dt, \quad \text{and} \quad C = \int e^{pt} \cos kt \, dt,$$

where p and k are constants.

To tackle these formidable-looking integrals, we resort to the device of integrating by parts, the general formula of which is

$$\int u \, dv = uv - \int v \, du$$

For this purpose, write $u = e^{pt}$ and $dv = \sin kt \, dt$ in S; then $du = pe^{pt} \, dt$, and $v = \int \sin kt \, dt = -\dfrac{1}{k} \cos kt$, omitting temporarily the constant.

Inserting these values, the integral S becomes

$$S = \int e^{pt} \sin kt \, dt = -\frac{1}{k} e^{pt} \cos kt - \int -\frac{1}{k} \cos kt \, pe^{pt} \, dt$$

$$= -\frac{1}{k} e^{pt} \cos kt + \frac{p}{k} \int e^{pt} \cos kt \, dt$$

$$= -\frac{1}{k} e^{pt} \cos kt + \frac{p}{k} C \quad \ldots\ldots\ldots\ldots\ldots\text{(i)}$$

Thus the dodge of integrating by parts turns S partly into C. But let us look at C. Writing $u = e^{pt}$, as before, $dv = \cos kt \, dt$, then $v = \dfrac{1}{k} \sin kt$; hence, the rule for integrating by parts gives

$$C = \int e^{pt} \cos kt \, dt = \frac{1}{k} e^{pt} \sin kt - \frac{p}{k} \int e^{pt} \sin kt \, dt$$

$$= \frac{1}{k} e^{pt} \sin kt - \frac{p}{k} S. \quad \ldots\ldots\ldots\ldots\text{(ii)}$$

The facts that S turns partly into C, and C partly into S might lead you to think that the integrals are intractable, but from the relations (i) and (ii), which may be regarded as two equations in S and C, the integrals themselves may be readily deduced.

Thus, substitute in (i) the value of C from (ii), then

$$S = -\frac{1}{k} e^{pt} \cos kt + \frac{p}{k} \left(\frac{1}{k} e^{pt} \sin kt - \frac{p}{k} S \right)$$

or $\qquad S\left(\frac{p^2}{k^2} + 1 \right) = \frac{1}{k^2} e^{pt} (p \sin kt - k \cos kt)$

from which $\qquad S = \frac{e^{pt}}{p^2 + k^2} (p \sin kt - k \cos kt)$

The integral C may be obtained in like manner by inserting in (ii) the equivalent of S given by (i); the final result is

$$C = \frac{e^{pt}}{p^2 + k^2} (p \cos kt + k \sin kt)$$

We have, therefore, the following very important integrals to add to our list, namely:

$$\int e^{pt} \sin kt \, dt = \frac{e^{pt}}{p^2 + k^2} (p \sin kt - k \cos kt) + E$$

$$\int e^{pt} \cos kt \, dt = \frac{e^{pt}}{p^2 + k^2} (p \cos kt + k \sin kt) + F$$

where E and F are the constants of integration.

Example 3.

Let $\qquad ay + b\frac{dy}{dt} = g \sin 2\pi nt$

First divide through by b.

$$\frac{dy}{dt} + \frac{a}{b}y = \frac{g}{b} \sin 2\pi nt$$

Now, as it stands, the left side is not integrable. But it can be made so by the artifice—and this is where skill and practice suggest a plan—of multiplying all the terms by $e^{\frac{a}{b}t}$, giving us:

$$\frac{dy}{dt}e^{\frac{a}{b}t} + \frac{a}{b}ye^{\frac{a}{b}t} = \frac{g}{b}e^{\frac{a}{b}t}\sin 2\pi nt$$

For if $u = ye^{\frac{a}{b}t}$, $\dfrac{du}{dt} = \dfrac{dy}{dt}e^{\frac{a}{b}t} + \dfrac{a}{b}ye^{\frac{a}{b}t}$

The equation thus becomes

$$\frac{du}{dt} = \frac{g}{b}e^{\frac{a}{b}t}\sin 2\pi nt$$

Hence, integrating gives

$$u \text{ or } ye^{\frac{a}{b}t} = \frac{g}{b}\int e^{\frac{a}{b}t}\sin 2\pi nt \, dt + K$$

But the right-hand integral is of the same form as S which has just been evaluated; hence putting $p = \dfrac{a}{b}$ and $k = 2\pi n$;

$$ye^{\frac{a}{b}t} = \frac{ge^{\frac{a}{b}t}}{a^2 + 4\pi^2 n^2 b^2}(a\sin 2\pi nt - 2\pi nb \cos 2\pi nt) + K$$

or

$$y = g\left\{\frac{a\sin 2\pi nt - 2\pi nb \cos 2\pi nt}{a^2 + 4\pi^2 n^2 b^2}\right\} + Ke^{-\frac{a}{b}t}$$

To simplify still further, let us imagine an angle ϕ such that $\tan \phi = 2\pi nb/a$.

Then $\sin \phi = \dfrac{2\pi nb}{\sqrt{a^2 + 4\pi^2 n^2 b^2}}$, and $\cos \phi = $

$\dfrac{a}{\sqrt{a^2 + 4\pi^2 n^2 b^2}}$. Substituting these, we get:

$$y = g\frac{\cos \phi \sin 2\pi nt - \sin \phi \cos 2\pi nt}{\sqrt{a^2 + 4\pi^2 n^2 b^2}}$$

or
$$y = g\frac{\sin\,(2\pi nt - \phi)}{\sqrt{a^2 + 4\pi^2 n^2 b^2}}$$

which is *the solution* desired, omitting the constant which dies out.

This is indeed none other than the equation of an alternating electric current, where g represents the amplitude of the electromotive force, n the frequency, a the resistance, b the coefficient of induction of the circuit, and ϕ is the phase angle of lag.

Example 4.

Suppose that $M\,dx + N\,dy = 0$

We could integrate this expression directly, if M were a function of x only, and N a function of y only; but, if both M and N are functions that depend on both x and y, how are we to integrate it? Is it itself an exact differential? That is: have M and N each been formed by partial differentiations from some common function U, or not? If they have, then

$$\frac{\partial U}{\partial x} = M, \text{ and } \frac{\partial U}{\partial y} = N$$

And if such a common function exists, then

$$\frac{\partial U}{\partial x}dx + \frac{\partial U}{\partial y}dy$$

is an exact differential.

Now the test of the matter is this. If the expression is an exact differential, it must be true that

$$\frac{\delta M}{\delta y} = \frac{\delta N}{\delta x}$$

for then $$\frac{\delta(\delta U)}{\delta x\,\delta y} = \frac{\delta(\delta U)}{\delta y\,\delta x}$$

which is necessarily true.

Take as an illustration the equation

$$(1 + 3xy)dx + x^2 dy = 0$$

Is this an exact differential or not? Apply the test.

$$\frac{\delta(1 + 3xy)}{\delta y} = 3x \qquad \frac{\delta(x^2)}{\delta x} = 2x$$

which do not agree. Therefore, it is not an exact differential, and the two functions $1 + 3xy$ and x^2 have not come from a common original function.

It is possible in such cases to discover, however, *an integrating factor*, that is to say, a factor such that if both are multiplied by this factor, the expression will become an exact differential. There is no one rule for discovering such an integrating factor; but experience will usually suggest one. In the present instance $2x$ will act as such. Multiplying by $2x$, we get

$$(2x + 6x^2y)dx + 2x^3dy = 0$$

Now apply the test to this.

$$\frac{\delta(2x + 6x^2y)}{\delta y} = 6x^2 \qquad \frac{\delta(2x^3)}{\delta x} = 6x^2$$

which agrees. Hence this is an exact derivative, and may be integrated. Now, if $w = 2x^3y$,

$$dw = 6x^2y\ dx + 2x^3dy$$

Hence $\qquad \displaystyle\int 6x^2y\ dx + \int 2x^3\ dy = w = 2x^3y$

so that we get $\qquad U = x^2 + 2x^3y + C$

Example 5. Let $\dfrac{d^2y}{dt^2} + n^2 y = 0$

In this case we have a differential equation of the second degree, in which y appears in the form of a second derivative, as well as in person. Transposing, we have

$$\frac{d^2y}{dt^2} = -n^2y$$

It appears from this that we have to do with a function such that its second derivative is proportional to itself, but with reversed sign. In Chapter 15 we found that there was such a function—namely, the *sine* (or the *cosine* also) which possessed this property. So, without further ado, we may guess that the solution will be of the form

$$y = A \sin (pt + q)$$

However, let us go to work.

Multiply both sides of the original equation by $2\dfrac{dy}{dt}$ and integrate, giving us $2\dfrac{d^2y}{dt^2}\dfrac{dy}{dt} + 2n^2 y \dfrac{dy}{dt} = 0$, and, as

$$2\frac{d^2y}{dt^2}\frac{dy}{dt} = \frac{d\left(\dfrac{dy}{dt}\right)^2}{dt}, \quad \left(\frac{dy}{dt}\right)^2 + n^2(y^2 - C^2) = 0$$

C being a constant. Then, taking the square roots,

$$\frac{dy}{dt} = n\sqrt{C^2 - y^2} \quad \text{and} \quad \frac{dy}{\sqrt{C^2 - y^2}} = n \cdot dt$$

But it can be shown that

$$\frac{1}{\sqrt{C^2 - y^2}} = \frac{d\left(\arcsin\dfrac{y}{C}\right)}{dy}$$

whence, passing from angles to sines,

$$\arcsin \frac{y}{C} = nt + C_1 \text{ and } y = C \sin (nt + C_1)$$

where C_1 is a constant angle that comes in by integration.
Or, preferably, this may be written

$$y = A \sin nt + B \cos nt, \text{ which is the solution.}$$

Example 6. $$\frac{d^2y}{dx^2} - n^2y = 0$$

Here we have obviously to deal with a function y which is such that its second derivative is proportional to itself. The only function we know that has this property is the exponential function, and we may be certain therefore that the solution of the equation will be of that form.

Proceeding as before, by multiplying through by $2\frac{dy}{dx}$, and integrating, we get $2\frac{d^2y}{dx^2}\frac{dy}{dx} - 2n^2y\frac{dy}{dx} = 0$, and, as

$$2\frac{d^2y}{dx^2}\frac{dy}{dx} = \frac{d\left(\frac{dy}{dx}\right)^2}{dx}, \quad \left(\frac{dy}{dx}\right)^2 - n^2(y^2 + c^2) = 0$$

$$\frac{dy}{dx} - n\sqrt{y^2 + c^2} = 0$$

where c is a constant, and $\dfrac{dy}{\sqrt{y^2 + c^2}} = n\, dx$.

To integrate this equation it is simpler to use hyperbolic functions.

Let $\quad\quad y = c \sinh u$, then $dy = c \cosh u\, du$, and

$$y^2 + c^2 = c^2 (\sinh^2 u + 1) = c^2 \cosh^2 u.$$

Therefore $\displaystyle\int \frac{dy}{\sqrt{y^2 + c^2}} = \int \frac{c \cosh u\, du}{c \cosh u} = \int du = u$

Hence, the integral of the equation

$$n\int dx = \int \frac{dy}{\sqrt{y^2 + c^2}}$$

is $\quad\quad nx + K = u$

where K is the constant of integration, and $c \sinh u = y$.

$$\sinh (nx + K) = \sinh u = \frac{y}{c}$$

or
$$y = c \sinh (nx + K)$$
$$= \tfrac{1}{2}c(e^{nx+K} - e^{-nx-K})$$
$$= Ae^{nx} + Be^{-nx}$$

where $A = \tfrac{1}{2}ce^{K}$ and $B = -\tfrac{1}{2}ce^{-K}$.

This solution which at first sight does not look as if it had any-thing to do with the original equation, shows that y consists of two terms, one of which grows exponentially as x increases, while the other term dies away as x increases.

Example 7.

Let
$$b\frac{d^2y}{dt^2} + a\frac{dy}{dt} + gy = 0.$$

Examination of this expression will show that, if $b = 0$, it has the form of Example 1, the solution of which was a negative ex-ponential. On the other hand, if $a = 0$, its form becomes the same as that of Example 6, the solution of which is the sum of a posi-tive and a negative exponential. It is therefore not very surprising to find that the solution of the present example is

$$y = (e^{-mt})(Ae^{nt} + Be^{-nt})$$

where
$$m = \frac{a}{2b} \quad \text{and} \quad n = \frac{\sqrt{a^2 - 4bg}}{2b}$$

The steps by which this solution is reached are not given here; they may be found in advanced treatises.

Example 8.

$$\frac{\delta^2 y}{\delta t^2} = a^2 \frac{\delta^2 y}{\delta x^2}$$

It was seen earlier that this equation was derived from the original

257

$$y = F(x + at) + f(x - at)$$

where F and f were any arbitrary functions of t.

Another way of dealing with it is to transform it by a change of variables into

$$\frac{\delta^2 y}{\delta u \cdot \delta v} = 0$$

where $u = x + at$, and $v = x - at$, leading to the same general solution. If we consider a case in which F vanishes, then we have simply

$$y = f(x - at)$$

and this merely states that, at the time $t = 0$, y is a particular function of x, and may be looked upon as denoting that the curve of the relation of y to x has a particular shape. Then any change in the value of t is equivalent simply to an alteration in the origin from which x is reckoned. That is to say, it indicates that, the form of the function being conserved, it is propagated along the x direction with a uniform velocity a; so that whatever the value of the ordinate y at any particular time t_0 at any particular point x_0, the same value of y will appear at the subsequent time t_1 at a point further along, the abscissa of which is $x_0 + a(t_1 - t_0)$. In this case the simplified equation represents the propagation of a wave (of any form) at a uniform speed along the x direction.

If the differential equation had been written

$$m\frac{d^2 y}{dt^2} = k\frac{d^2 y}{dx^2}$$

the solution would have been the same, but the velocity of propagation would have had the value

$$a = \sqrt{\frac{k}{m}}$$

Exercises XX

(See answers on page 308)

Try to solve the following equations.

(1) $\dfrac{dT}{d\theta} = \mu T$, given that μ is constant, and when $\theta = 0$, $T = T_0$.

(2) $\dfrac{d^2s}{dt^2} = a$, where a is constant. When $t = 0$, $s = 0$ and $\dfrac{ds}{dt} = u$.

(3) $\dfrac{di}{dt} + 2i = \sin 3t$, it being known that $i = 0$ when $t = 0$.

(*Hint.* Multiply out by e^{2t}.)

CHAPTER 22

A Little More About Curvature of Curves

In Chapter 12 we have learned how we can find out which way a curve is curved, that is, whether it curves upwards or downwards towards the right. This gave us no indication whatever as to *how much* the curve is curved, or, in other words, what is its *curvature*.

By *curvature* of a line, we mean the amount of bending or deflection taking place along a certain length of the line, say along a portion of the line the length of which is one unit of length (the same unit which is used to measure the radius, whether it be one inch, one foot, or any other unit). For instance, consider two circular paths of center O or O' and of equal lengths AB, $A'B'$ (see Fig 64). When pasing from A to B along thr arc AB of the first one, one changes one's direction from AP to BQ, since at A

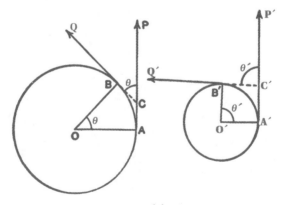

FIG. 64.

one faces in the direction AP and at B one faces in the direction BQ. In other words, in walking from A to B one unconsciously turns round through the angle PCQ, which is equal to the angle AOB. Similarly, in passing from A' to B', along the arc $A'B'$, of equal length to AB, on the second path, one turns round through the angle $P'C'Q'$, which is equal to the angle $A'O'B'$, obviously, *greater* than the corresponding angle AOB. The second path bends therefore more than the first for an equal length.

This fact is expressed by saying that the *curvature* of the second path is greater than that of the first one. The larger the circle, the lesser the bending, that is, the lesser the curvature. If the radius of the first circle is 2, 3, 4, . . . etc. times greater than the radius of the second, then the angle of bending or deflection along an arc of unit length will be 2, 3, 4, . . . etc. times less on the first circle than on the second, that is, it will be $\frac{1}{2}, \frac{1}{3}, \frac{1}{4}, \ldots$. etc. of the bending or deflection along the arc of same length on the second circle. In other words, the *curvature* of the first circle will be $\frac{1}{2}, \frac{1}{3}, \frac{1}{4}, \ldots$ etc. of that of the second circle. We see that, as the radius becomes 2, 3, 4, . . . etc. times greater, the curvature becomes 2, 3, 4, . . . etc. times smaller, and this is expressed by saying that *the curvature of a circle is inversely proportional to the radius of the circle*, or

$$\text{curvature} = k \times \frac{1}{\text{radius}}$$

where k is a constant. It is agreed to take $k = 1$, so that

$$\text{curvature} = \frac{1}{\text{radius}}$$

always.

If the radius becomes infinitely great, the curvature becomes $\frac{1}{\text{infinity}} = \text{zero}$, since when the denominator of a fraction is infinitely large, the value of the fraction is infinitely small. For this reason mathematicians sometimes consider a straight line as an arc of circle of infinite radius, or zero curvature.

In the case of a circle, which is perfectly symmetrical and uniform, so that the curvature is the same at every point of its circumference, the above method of expressing the curvature is perfectly definite. In the case of any other curve, however, the curvature is not the same at different points, and it may differ considerably even for two points fairly close to one another. It would not then be accurate to take the amount of bending or deflection between two points as a measure of the curvature of the arc between these points, unless this arc is very small, in fact, unless it is infinitesimal small.

If then we consider a very small arc such as *AB* (see Fig. 65), and if we draw such a circle that an arc *AB* of this circle coincides with the arc *AB* of the curve more closely than would be the case with any other circle, then the curvature of this circle may be taken as the curvature of the arc *AB* of the curve. The smaller the arc *AB*, the easier it will be to find a circle an arc of which most nearly coincides with the arc *AB* of the curve. When *A* and *B* are very near one another, so that *AB* is so small so that the length *ds* of the arc *AB* is practically negligible, then the coincidence of the two arcs, of circle and of curve, may be considered as being practically perfect, and the curvature of the curve at the point *A* (or *B*), being then the same as the curvature of the circle, will be expressed by the reciprocal of the radius of this circle, that is, by $\dfrac{1}{OA}$, according to our way of measuring curvature, explained above.

Now, at first, you may think that, if *AB* is very small, then the circle must be very small also. A little thinking will, however,

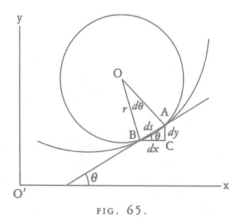

FIG. 65.

cause you to perceive that it is by no means necessarily so, and that the circle may have any size, according to the amount of bending of the curve along this very small arc AB. In fact, if the curve is almost flat at that point, the circle will be extremely large. This circle is called the *circle of curvature*, or the *osculating circle* at the point considered. Its radius is the *radius of curvature* of the curve at that particular point.

If the arc AB is represented by ds and the angle AOB by $d\theta$, then, if r is the radius of curvature,

$$ds = r\, d\theta \quad \text{or} \quad \frac{d\theta}{ds} = \frac{1}{r}$$

The secant AB makes with the axis OX the angle θ, and it will be seen from the small triangle ABC that $\dfrac{dy}{dx} = \tan\theta$. When AB is infinitesimal small, so that B practically coincides with A, the line AB becomes a tangent to the curve at the point A (or B).

Now, $\tan\theta$ depends on the position of the point A (or B, which is supposed to nearly coincide with it), that is, it depends on x, or, in other words, $\tan\theta$ is "a function" of x.

Differentiating with regard to x to get the slope, we get

$$\frac{d\left(\dfrac{dy}{dx}\right)}{dx} = \frac{d(\tan\theta)}{dx} \quad \text{or} \quad \frac{d^2y}{dx^2} = \sec^2\theta\,\frac{d\theta}{dx} = \frac{1}{\cos^2\theta}\frac{d\theta}{dx}$$

hence

$$\frac{d\theta}{dx} = \cos^2\theta\,\frac{d^2y}{dx^2}$$

But $\dfrac{dx}{ds} = \cos\theta$, and for $\dfrac{d\theta}{ds}$ one may write $\dfrac{d\theta}{dx} \times \dfrac{dx}{ds}$; therefore

$$\frac{1}{r} = \frac{d\theta}{ds} = \frac{d\theta}{dx} \times \frac{dx}{ds} = \cos^3\theta\,\frac{d^2y}{dx^2} = \frac{\dfrac{d^2y}{dx^2}}{\sec^3\theta}$$

but $\sec\theta = \sqrt{1 + \tan^2\theta}$; hence

263

$$\frac{1}{r} = \frac{\dfrac{d^2y}{dx^2}}{\left(\sqrt{1 + \tan^2 \theta}\right)^3} = \frac{\dfrac{d^2y}{dx^2}}{\left\{1 + \left(\dfrac{dy}{dx}\right)^2\right\}^{\frac{3}{2}}}$$

and finally,

$$r = \frac{\left\{1 + \left(\dfrac{dy}{dx}\right)^2\right\}^{\frac{3}{2}}}{\dfrac{d^2y}{dx^2}}$$

The numerator, being a square root, may have the sign + or the sign −. One must select for it the same sign as the denominator, so as to have *r* positive always, as a negative radius would have no meaning.

It has been shown (Chapter 12) that if $\dfrac{d^2y}{dx^2}$ is positive, the curve is concave upward, while if $\dfrac{d^2y}{dx^2}$ is negative, the curve is concave downward. If $\dfrac{d^2y}{dx^2} = 0$, the radius of curvature is infinitely great, that is, the corresponding portion of the curve is a bit of straight line. This necessarily happens whenever a curve gradually changes from being convex to concave to the axis of *x* or vice versa. The point where this occurs is called a *point of inflection.*

The center of the circle of curvature is called the *center of curvature.* If its coordinates are x_1, y_1, then the equation of the circle is

$$(x - x_1)^2 + (y - y_1)^2 = r^2$$

hence

$$2(x - x_1)dx + 2(y - y_1)dy = 0$$

and

$$x - x_1 + (y - y_1)\frac{dy}{dx} = 0 \quad \text{.....................(1)}$$

Why did we differentiate? To get rid of the constant r. This leaves but two unknown constants x_1 and y_1; differentiate again; you shall get rid of one of them. This last differentiation is not quite as easy as it seems: let us do it together; we have:

$$\frac{d(x)}{dx} + \frac{d\left[(y - y_1)\dfrac{dy}{dx}\right]}{dx} = 0$$

the numerator of the second term is a product; hence differentiating it gives

$$(y - y_1)\frac{d\left(\dfrac{dy}{dx}\right)}{dx} + \frac{dy}{dx}\frac{d(y - y_1)}{dx} = (y - y_1)\frac{d^2y}{dx^2} + \left(\frac{dy}{dx}\right)^2$$

so that the result of differentiating (1) is

$$1 + \left(\frac{dy}{dx}\right)^2 + (y - y_1)\frac{d^2y}{dx^2} = 0$$

from this we at once get

$$y_1 = y + \frac{1 + \left(\dfrac{dy}{dx}\right)^2}{\dfrac{d^2y}{dx^2}}$$

Replacing in (1), we get

$$(x - x_1) + \left\{y - y - \frac{1 + \left(\dfrac{dy}{dx}\right)^2}{\dfrac{d^2y}{dx^2}}\right\}\frac{dy}{dx} = 0$$

and finally,
$$x_1 = x - \frac{\dfrac{dy}{dx}\left\{1 + \left(\dfrac{dy}{dx}\right)^2\right\}}{\dfrac{d^2y}{dx^2}}$$

x_1 and y_1 give the position of the center of curvature. The use of these formulae will be best seen by carefully going through a few worked-out examples.

Example 1.   Find the radius of curvature and the coordinates of the center of curvature of the curve $y = 2x^2 - x + 3$ at the point $x = 0$.

We have
$$\frac{dy}{dx} = 4x - 1 \qquad \frac{d^2y}{dx^2} = 4$$

$$r = \frac{\pm\left\{1 + \left(\dfrac{dy}{dx}\right)^2\right\}^{\frac{3}{2}}}{\dfrac{d^2y}{dx^2}} = \frac{\{1 + (4x - 1)^2\}^{\frac{3}{2}}}{4}$$

when $x = 0$; this becomes
$$\frac{\{1 + (-1)^2\}^{\frac{3}{2}}}{4} = \frac{\sqrt{8}}{4} = 0.707$$

If x_1, y_1 are the coordinates of the center of curvature, then

$$x_1 = x - \frac{\dfrac{dy}{dx}\left\{1 + \left(\dfrac{dy}{dx}\right)^2\right\}}{\dfrac{d^2y}{dx^2}} = x - \frac{(4x - 1)\{1 + (4x - 1)^2\}}{4}$$

$$= 0 - \frac{(-1)\{1 + (-1)^2\}}{4} = \frac{1}{2}$$

when $x = 0$, $y = 3$, so that

$$y_1 = y + \frac{1 + \left(\dfrac{dy}{dx}\right)^2}{\dfrac{d^2y}{dx^2}} = y + \frac{1 + (4x - 1)^2}{4} = 3 + \frac{1 + (-1)^2}{4} = 3\tfrac{1}{2}$$

Plot the curve and draw the circle; it is both interesting and instructive. The values can be checked easily, as since when $x = 0$, $y = 3$, here

$$x_1^2 + (y_1 - 3)^2 = r^2 \quad \text{or} \quad 0.5^2 + 0.5^2 = 0.5 = 0.707^2$$

266

Example 2. Find the radius of curvature and the position of the center of curvature of the curve $y^2 = mx$ at the point for which $y = 0$.

Here
$$y = m^{\frac{1}{2}}x^{\frac{1}{2}}, \quad \frac{dy}{dx} = \frac{1}{2}m^{\frac{1}{2}}x^{-\frac{1}{2}} = \frac{m^{\frac{1}{2}}}{2x^{\frac{1}{2}}}$$

$$\frac{d^2y}{dx^2} = -\frac{1}{2} \times \frac{m^{\frac{1}{2}}}{2}x^{-\frac{3}{2}} = -\frac{m^{\frac{1}{2}}}{4x^{\frac{3}{2}}}$$

hence
$$\frac{\pm\left\{1 + \left(\dfrac{dy}{dx}\right)^2\right\}^{\frac{3}{2}}}{\dfrac{d^2y}{dx^2}} = \frac{\pm\left\{1 + \dfrac{m}{4x}\right\}^{\frac{3}{2}}}{-\dfrac{m^{\frac{1}{2}}}{4x^{\frac{3}{2}}}} = \frac{(4x + m)^{\frac{3}{2}}}{2m^{\frac{1}{2}}}$$

taking the $-$ sign at the numerator, so as to have r positive.

Since, when $y = 0$, $x = 0$, we get $r = \dfrac{m^{\frac{3}{2}}}{2m^{\frac{1}{2}}} = \dfrac{m}{2}$

Also, if x_1, y_1 are the coordinates of the center,

$$x_1 = x - \frac{\dfrac{dy}{dx}\left\{1 + \left(\dfrac{dy}{dx}\right)^2\right\}}{\dfrac{d^2y}{d^2x}} = x - \frac{\dfrac{m^{\frac{1}{2}}}{2x^{\frac{1}{2}}}\left\{1 + \dfrac{m}{4x}\right\}}{-\dfrac{m^{\frac{1}{2}}}{4x^{\frac{3}{2}}}}$$

$$= x + \frac{4x + m}{2} = 3x + \frac{m}{2}$$

when $x = 0$, then $x_1 = \dfrac{m}{2}$

Also
$$y_1 = y + \frac{1 + \left(\dfrac{dy}{dx}\right)^2}{\dfrac{d^2y}{dx^2}} = m^{\frac{1}{2}}x^{\frac{1}{2}} - \frac{1 + \dfrac{m}{4x}}{\dfrac{m^{\frac{1}{2}}}{4x^{\frac{3}{2}}}} = -\frac{4x^{\frac{3}{2}}}{m^{\frac{1}{2}}}$$

when $x = 0$, $y_1 = 0$.

Example 3. Show that the circle is a curve of constant curvature.

If x_1, y_1 are the coordinates of the center, and R is the radius, the equation of the circle in rectangular coordinates is

$$(x - x_1)^2 + (y - y_1)^2 = R^2$$

Let $x - x_1 = R \cos \theta$, then

$$(y - y_1)^2 = R^2 - R^2 \cos^2\theta = R^2(1 - \cos^2\theta) = R^2 \sin^2\theta$$

$$y - y_1 = R \sin \theta$$

R, θ are thus the polar coordinates of any point on the circle referred to its center as pole.

Since $x - x_1 = R \cos \theta$, and $y - y_1 = R \sin \theta$,

$$\frac{dx}{d\theta} = -R \sin \theta, \quad \frac{dy}{d\theta} = R \cos \theta$$

$$\frac{dy}{dx} = \frac{dy}{d\theta} \cdot \frac{d\theta}{dx} = -\cot \theta$$

Further, $\quad \dfrac{d^2y}{dx^2} = -(-\csc^2\theta)\dfrac{d\theta}{dx} = \csc^2\theta \cdot \left(-\dfrac{\csc \theta}{R}\right)$

$$= -\frac{\csc^3 \theta}{R}. \quad \text{(See Ex. 5, Chapter 15.)}$$

Hence $\quad r = \dfrac{\pm(1 + \cot^2 \theta)^{\frac{3}{2}}}{-\dfrac{\csc^3\theta}{R}} = \dfrac{R \csc^3\theta}{\csc^3\theta} = R$

Thus the radius of curvature is constant and equal to the radius of the circle.

Example 4. Find the radius of curvature of the curve $x = 2 \cos^3 t$, $y = 2 \sin^3 t$ at any point (x, y).

Here $\quad dx = -6 \cos^2 t \sin t\, dt$ (see Ex. 2, Chapter 15)

and $\quad dy = 6 \sin^2 t \cos t\, dt.$

$$\frac{dy}{dx} = -\frac{6 \sin^2 t \cos t\, dt}{6 \sin t \cos^2 t\, dt} = -\frac{\sin t}{\cos t} = -\tan t$$

Hence $\dfrac{d^2y}{dx^2} = \dfrac{d}{dt}(-\tan t)\dfrac{dt}{dx} = \dfrac{-\sec^2 t}{-6\cos^2 t \sin t} = \dfrac{\sec^4 t}{6\sin t}$

$$r = \frac{\pm(1 + \tan^2 t)^{\frac{3}{2}} \times 6\sin t}{\sec^4 t} = \frac{6\sec^3 t \sin t}{\sec^4 t}$$

$$= 6\sin t \cos t = 3\sin 2t, \text{ for } 2\sin t \cos t = \sin 2t$$

Example 5. Find the radius and the center of curvature of the curve $y = x^3 - 2x^2 + x - 1$ at points where $x = 0$, $x = 0.5$ and $x = 1$. Find also the position of the point of inflection of the curve.

Here $\dfrac{dy}{dx} = 3x^2 - 4x + 1$, $\dfrac{d^2y}{dx^2} = 6x - 4$

$$r = \frac{\{1 + (3x^2 - 4x + 1)^2\}^{\frac{3}{2}}}{6x - 4}$$

$$x_1 = x - \frac{(3x^2 - 4x + 1)\{1 + (3x^2 - 4x + 1)^2\}}{6x - 4}$$

$$y_1 = y + \frac{1 + (3x^2 - 4x + 1)^2}{6x - 4}$$

When $x = 0$, $y = -1$,

$$r = \frac{\sqrt{8}}{4} = 0.707, x_1 = 0 + \tfrac{1}{2} = 0.5, y_1 = -1 - \tfrac{1}{2} = -1.5.$$

When $x = 0.5$, $y = -0.875$:

$$r = \frac{-\{1 + (-0.25)^2\}^{\frac{3}{2}}}{-1} = 1.09$$

$$x_1 = 0.5 - \frac{-0.25 \times 1.0625}{-1} = 0.23$$

$$y_1 = -0.875 + \frac{1.0625}{-1} = -1.94$$

269

When $x = 1$, $y = -1$.

$$r = \frac{(1 + 0)^{\frac{3}{2}}}{2} = 0.5$$

$$x_1 = 1 - \frac{0 \times (1 + 0)}{2} = 1$$

$$y_1 = -1 + \frac{1 + 0^2}{2} = -0.5$$

At the point of inflection $\dfrac{d^2y}{dx^2} = 0$, $6x - 4 = 0$, and $x = \frac{2}{3}$; hence

$y = -0.926$.

Example 6. Find the radius and center of curvature of the curve $y = \dfrac{a}{2}\left\{e^{\frac{x}{a}} + e^{-\frac{x}{a}}\right\}$, at the point for which $x = 0$. (This curve is called

the *catenary*, as a hanging chain affects the same slope exactly.)
The equation of the curve may be written

$$y = \frac{a}{2} e^{\frac{x}{a}} + \frac{a}{2} e^{-\frac{x}{a}}$$

then,

$$\frac{dy}{dx} = \frac{a}{2} \times \frac{1}{a} e^{\frac{x}{a}} - \frac{a}{2} \times \frac{1}{a} e^{-\frac{x}{a}} = \frac{1}{2}\left(e^{\frac{x}{a}} - e^{-\frac{x}{a}}\right)$$

Similarly

$$\frac{d^2y}{dx^2} = \frac{1}{2a}\left\{e^{\frac{x}{a}} + e^{-\frac{x}{a}}\right\} = \frac{1}{2a} \times \frac{2y}{a} = \frac{y}{a^2}$$

$$r = \frac{\left\{1 + \frac{1}{4}\left(e^{\frac{x}{a}} - e^{-\frac{x}{a}}\right)^2\right\}^{\frac{3}{2}}}{\dfrac{y}{a^2}} = \frac{a^2}{8y}\sqrt{\left(2 + e^{\frac{2x}{a}} + e^{-\frac{2x}{a}}\right)^3}$$

since $e^{\frac{x}{a}-\frac{x}{a}} = e^0 = 1$, or

$$r = \frac{a^2}{8y}\sqrt{\left(2e^{\frac{x}{a}-\frac{x}{a}} + e^{\frac{2x}{a}} + e^{-\frac{2x}{a}}\right)^3} = \frac{a^2}{8y}\sqrt{\left(e^{\frac{x}{a}} + e^{-\frac{x}{a}}\right)^6} = \frac{y^2}{a}$$

when $\qquad x = 0, \; y = \frac{a}{2}(e^0 + e^0) = a$

hence $\qquad\qquad\qquad r = \frac{a^2}{a} = a$

The radius of curvature at the vertex is equal to the constant a.

Also when $\quad x = 0, \; x_1 = 0 - \dfrac{0(1+0)}{\dfrac{1}{a}} = 0$

and $\qquad\qquad y_1 = y + \dfrac{1+0}{\dfrac{1}{a}} = a + a = 2a$

As defined previously

$$\frac{1}{2}\left(e^{\frac{x}{a}} + e^{-\frac{x}{a}}\right) = \cosh\frac{x}{a}$$

thus the equation of the catenary may be written in the form

$$y = a\cosh\frac{x}{a}$$

It will therefore be a useful exercise for you to verify the above results from this form of the equation.

You are now sufficiently familiar with this type of problem to work out the following exercises by yourself. You are advised to check your answers by careful plotting of the curve and construction of the circle of curvature, as explained in Example 4.

Exercises XXI

(See answers on page 308)

(1) Find the radius of curvature and the position of the center of curvature of the curve $y = e^x$ at the point for which $x = 0$.

271

(2) Find the radius and the center of curvature of the curve $y = x\left(\dfrac{x}{2} - 1\right)$ at the point for which $x = 2$.

(3) Find the point or points of curvature unity in the curve $y = x^2$.

(4) Find the radius and the center of curvature of the curve $xy = m$ at the point for which $x = \sqrt{m}$.

(5) Find the radius and the center of curvature of the curve $y^2 = 4ax$ at the point for which $x = 0$.

(6) Find the radius and the center of curvature of the curve $y = x^3$ at the points for which $x = \pm 0.9$ and also $x = 0$.

(7) Find the radius of curvature and the coordinates of the center of curvature of the curve

$$y = x^2 - x + 2$$

at the two points for which $x = 0$ and $x = 1$, respectively. Find also the maximum or minimum value of y. Verify graphically all your results.

(8) Find the radius of curvature and the coordinates of the center of curvature of the curve

$$y = x^3 - x - 1$$

at the points for which $x = -2$, $x = 0$, and $x = 1$.

(9) Find the coordinates of the point or points of inflection of the curve $y = x^3 + x^2 + 1$.

(10) Find the radius of curvature and the coordinates of the center of curvature of the curve

$$y = (4x - x^2 - 3)^{\frac{1}{2}}$$

at the points for which $x = 1.2$, $x = 2$ and $x = 2.5$. What is this curve?

(11) Find the radius and the center of curvature of the curve $y = x^3 - 3x^2 + 2x + 1$ at the points for which $x = 0$, $x = +1.5$. Find also the position of the point of inflection.

(12) Find the radius and center of curvature of the curve $y = \sin\theta$ at the points for which $\theta = \dfrac{\pi}{4}$ and $\theta = \dfrac{\pi}{2}$. Find the position of the points of inflection.

(13) Draw a circle of radius 3, the center of which has for its

coordinates $x = 1$, $y = 0$. Deduce the equation of such a circle from first principles. Find by calculation the radius of curvature and the coordinates of the center of curvature for several suitable points, as accurately as possible, and verify that you get the known values.

(14) Find the radius and center of curvature of the curve $y = \cos \theta$ at the points for which $\theta = 0$, $\theta = \dfrac{\pi}{4}$ and $\theta = \dfrac{\pi}{2}$.

(15) Find the radius of curvature and the center of curvature of the ellipse $\dfrac{x^2}{a^2} + \dfrac{y^2}{b^2} = 1$ at the points for which $x = 0$ and at the points for which $y = 0$.

(16) When a curve is defined by equations in the form

$$x = F(\theta), \quad y = f(\theta)$$

the radius r of curvature is given by

$$r = \left\{ \left(\frac{dx}{d\theta} \right)^2 + \left(\frac{dy}{d\theta} \right)^2 \right\}^{\frac{3}{2}} \bigg/ \left(\frac{dx}{d\theta} \cdot \frac{d^2y}{d\theta^2} - \frac{dy}{d\theta} \cdot \frac{d^2x}{d\theta^2} \right)$$

Apply the formula to find r for the curve

$$x = a(\theta - \sin \theta), \quad y = a(1 - \cos \theta)$$

273

CHAPTER 23

How To Find the Length of an Arc on a Curve

Since an arc on any curve is made up of a lot of little bits of straight lines joined end to end, if we could add all these little bits, we would get the length of the arc. But we have seen that to add a lot of little bits together is precisely what is called integration, so that it is likely that, since we know how to integrate, we can find also the length of an arc on any curve, provided that the equation of the curve is such that it lends itself to integration.

If MN is an arc on any curve, the length s of which is required (see Fig. 66a), if we call "a little bit" of the arc ds, then we see at once that

$$(ds)^2 = (dx)^2 + (dy)^2$$

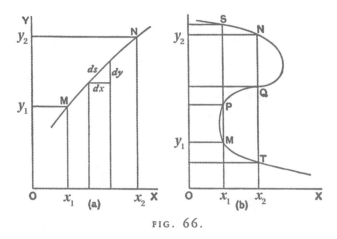

FIG. 66.

274

or either

$$ds = \sqrt{1 + \left(\frac{dx}{dy}\right)^2}\, dy \quad \text{or} \quad ds = \sqrt{1 + \left(\frac{dy}{dx}\right)^2}\, dx$$

Now the arc MN is made up of the sum of all the little bits ds between M and N, that is, between x_1 and x_2, or between y_1 and y_2, so that we get either

$$s = \int_{x_1}^{x_2} \sqrt{1 + \left(\frac{dy}{dx}\right)^2}\, dx \quad \text{or} \quad s = \int_{y_1}^{y_2} \sqrt{1 + \left(\frac{dx}{dy}\right)^2}\, dy$$

That is all!

The second integral is useful when there are several points of the curve corresponding to the given values of x (as in Fig. 66b). In this case the integral between x_1 and x_2 leaves a doubt as to the exact portion of the curve, the length of which is required. It may be ST, instead of MN, or SQ; by integrating between y_1 and y_2 the uncertainty is removed, and in this case one should use the second integral.

If instead of x and y coordinates—or Cartesian coordinates, as they are named from the French mathematician Descartes, who invented them—we have r and θ coordinates (or polar coordinates); then, if MN be a small arc of length ds on any curve, the length s of which is required (see Fig. 67), O being the pole, then the distance ON will generally differ from OM by a small amount dr. If the small angle MON is called $d\theta$, then, the polar coordinates of the point M being θ and r, those of N are $(\theta + d\theta)$ and $(r + dr)$. Let MP be perpendicular to ON, and let $OR = OM$; then

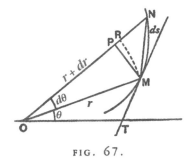

FIG. 67.

275

$RN = dr$, and this is very nearly the same as PN, as long as $d\theta$ is a very small angle. Also $RM = r\,d\theta$, and RM is very nearly equal to PM, and the arc MN is very nearly equal to the chord MN. In fact we can write $PN = dr$, $PM = r\,d\theta$, and arc $MN =$ chord MN without appreciable error, so that we have:

$$(ds)^2 = (\text{chord } MN)^2 = \overline{PN}^2 + \overline{PM}^2 = dr^2 + r^2 d\theta^2$$

Dividing by $d\theta^2$ we get $\left(\dfrac{ds}{d\theta}\right)^2 = r^2 + \left(\dfrac{dr}{d\theta}\right)^2$; hence

$$\frac{ds}{d\theta} = \sqrt{r^2 + \left(\frac{dr}{d\theta}\right)^2} \quad \text{and} \quad ds = \sqrt{r^2 + \left(\frac{dr}{d\theta}\right)^2}\, d\theta$$

hence, since the length s is made up of the sum of all the little bits ds, between values of $\theta = \theta_1$ and $\theta = \theta_2$, we have

$$s = \int_{\theta_1}^{\theta_2} ds = \int_{\theta_1}^{\theta_2} \sqrt{r^2 + \left(\frac{dr}{d\theta}\right)^2}\, d\theta$$

We can proceed at once to work out a few examples.

Example 1. The equation of a circle, the center of which is at the origin—or intersection of the axis of x with the axis of y—is $x^2 + y^2 = r^2$; find the length of an arc of one quadrant.

$$y^2 = r^2 - x^2 \text{ and } 2y\,dy = -2x\,dx, \text{ so that } \frac{dy}{dx} = -\frac{x}{y}$$

hence

$$s = \int \sqrt{\left[1 + \left(\frac{dy}{dx}\right)^2\right]}\,dx = \int \sqrt{\left(1 + \frac{x^2}{y^2}\right)}\,dx$$

and since $y^2 = r^2 - x^2$,

$$s = \int \sqrt{\left(1 + \frac{x^2}{r^2 - x^2}\right)}\,dx = \int \frac{r\,dx}{\sqrt{r^2 - x^2}}$$

The length we want—one quadrant—extends from a point for which $x = 0$ to another point for which $x = r$. We express this by writing

$$s = \int_{x=0}^{x=r} \frac{r\,dx}{\sqrt{r^2 - x^2}}$$

or, more simply, by writing

$$s = \int_0^r \frac{r\,dx}{\sqrt{r^2 - x^2}}$$

the 0 and r to the right of the sign of integration merely meaning that the integration is only to be performed on a portion of the curve, namely, that between $x = 0$, $x = r$, as we have seen.

Here is a fresh integral for you! Can you manage it?

In Chapter 15 we differentiated $y = \arcsin x$ and found $\dfrac{dy}{dx} = \dfrac{1}{\sqrt{1 - x^2}}$. If you have tried all sorts of variations of the given examples (as you ought to have done!), you perhaps tried to differentiate something like $y = a\,\arcsin \dfrac{x}{a}$, which gave

$$\frac{dy}{dx} = \frac{a}{\sqrt{a^2 - x^2}} \quad \text{or} \quad dy = \frac{a\,dx}{\sqrt{a^2 - x^2}}$$

that is, just the same expression as the one we have to integrate here.

Hence $s = \displaystyle\int \frac{r\,dx}{\sqrt{r^2 - x^2}} = r\,\arcsin \dfrac{x}{r} + C$, C being a constant.

As the integration is only to be made between $x = 0$ and $x = r$, we write

$$s = \int_0^r \frac{r\,dx}{\sqrt{r^2 - x^2}} = \left[r\,\arcsin \frac{x}{r} + C \right]_0^r$$

proceeding then as explained in Example (1), Chapter 14, we get

$$s = r\,\arcsin \frac{r}{r} + C - r\,\arcsin \frac{0}{r} - C, \quad \text{or} \quad s = r \times \frac{\pi}{2}$$

277

since arcsin 1 is 90° or $\dfrac{\pi}{2}$ and arc-sin 0 is zero, and the constant C disappears, as has been shown.

The length of the quadrant is therefore $\dfrac{\pi r}{2}$, and the length of the circumference, being four times this, is $4 \times \dfrac{\pi r}{2} = 2\pi r$.

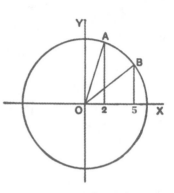

FIG. 68.

Example 2. Find the length of the arc AB between $x_1 = 2$ and $x_2 = 5$, in the circumference $x^2 + y^2 = 6^2$ (see Fig. 68).

Here, proceeding as in the previous example,

$$s = \left[\, r \arcsin \frac{x}{r} + C \,\right]_{x_1}^{x_2} = \left[\, 6 \arcsin \frac{x}{6} + C \,\right]_{2}^{5}$$

$$= 6\left[\, \arcsin \frac{5}{6} - \arcsin \frac{2}{6} \,\right] = 6(0.9851 - 0.3398)$$

$$= 3.8716 \text{ (the arcs being expressed in radians)}.$$

It is always well to check results obtained by a new and yet unfamiliar method. This is easy, for

$$\cos AOX = \tfrac{2}{6} = \tfrac{1}{3} \text{ and } \cos BOX = \tfrac{5}{6}$$

hence $AOB = AOX - BOX = \arccos \tfrac{1}{3} = \arccos \tfrac{5}{6} = 0.6453$ radians, and the length is $6 \times 0.6453 = 3.8716$.

Example 3. Find the length of an arc of the curve

$$y = \frac{a}{2}\left\{ e^{\frac{x}{a}} + e^{-\frac{x}{a}} \right\}$$

between $x = 0$ and $x = a$. (This curve is the *catenary*.)

278

$$y = \frac{a}{2}e^{\frac{x}{a}} + \frac{a}{2}e^{-\frac{x}{a}}, \frac{dy}{dx} = \frac{1}{2}\left\{e^{\frac{x}{a}} - e^{-\frac{x}{a}}\right\}$$

$$s = \int \sqrt{1 + \frac{1}{4}\left\{e^{\frac{x}{a}} - e^{-\frac{x}{a}}\right\}^2} \, dx$$

$$= \frac{1}{2}\int \sqrt{4 + e^{\frac{2x}{a}} + e^{-\frac{2x}{a}} - 2e^{\frac{x}{a}-\frac{x}{a}}} \, dx$$

Now

$$e^{\frac{x}{a}-\frac{x}{a}} = e^0 = 1, \text{ so that } s = \frac{1}{2}\int \sqrt{2 + e^{\frac{2x}{a}} + e^{-\frac{2x}{a}}} \, dx$$

we can replace 2 by $2 \times e^0 = 2 \times e^{\frac{x}{a}-\frac{x}{a}}$; then

$$s = \frac{1}{2}\int \sqrt{e^{\frac{2x}{a}} + 2e^{\frac{x}{a}-\frac{x}{a}} + e^{-\frac{2x}{a}}} \, dx$$

$$= \frac{1}{2}\int \sqrt{\left(e^{\frac{x}{a}} + e^{-\frac{x}{a}}\right)^2} \, dx = \frac{1}{2}\int \left(e^{\frac{x}{a}} + e^{-\frac{x}{a}}\right) dx$$

$$= \frac{1}{2}\int e^{\frac{x}{a}} \, dx + \frac{1}{2}\int e^{-\frac{x}{a}} \, dx = \frac{a}{2}\left[e^{\frac{x}{a}} - e^{-\frac{x}{a}}\right]$$

Here $\quad s = \frac{a}{2}\left[e^{\frac{x}{a}} - e^{-\frac{x}{a}}\right]_0^a = \frac{a}{2}[e^1 - e^{-1} - 1 + 1]$

and $\quad s = \frac{a}{2}\left(e - \frac{1}{e}\right) = 1.1752a.$

Example 4. A curve is such that the length of the tangent at any point P (see Fig. 69) from P to the intersection T of the tangent with a fixed line AB is a constant length a. Find an expression for

279

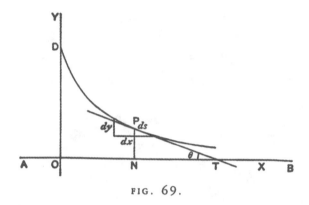

FIG. 69.

an arc of this curve—which is called the tractrix —and find the length, when $a = 3$, between the ordinates $y = a$ and $y = 1$.

We shall take the fixed line for the axis of x. The point D, with $DO = a$, is a point on the curve, which must be tangent to OD at D. We take OD as the axis of y. $PT = a$, $PN = y$, $ON = x$.

If we consider a small portion ds of the curve, at P, then $\sin \theta = \dfrac{dy}{ds} = -\dfrac{y}{a}$ (minus because the curve slopes *downwards* to the right.

Hence $\dfrac{ds}{dy} = -\dfrac{a}{y}$, $ds = -a\dfrac{dy}{y}$ and $s = -a\displaystyle\int\dfrac{dy}{y}$, that is,

$$s = -a \ln y + C$$

When $x = 0$, $s = 0$, $y = a$, so that $0 = -a \ln a + C$, and $C = a \ln a$.

It follows that $s = a \ln a - a \ln y = a \ln \dfrac{a}{y}$.

When $a = 3$, s between $y = a$ and $y = 1$ is therefore

$$s = 3\left[\ln \frac{3}{y} \right]_1^3 = 3(\ln 1 - \ln 3) = 3 \times (0 - 1.0986)$$

$$= -3.296 \text{ or } 3.296,$$

as the sign — refers merely to the direction in which the length was measured, from D to P, or from P to D.

Note that this result has been obtained without a knowledge of the equation of the curve. This is sometimes possible. In order to get the length of an arc between two points given by their abscissae, however, it is necessary to know the equation of the curve; this is easily obtained as follows:

$$\frac{dy}{dx} = -\tan\theta = -\frac{y}{\sqrt{a^2 - y^2}}, \text{ since } PT = a$$

hence $\quad dx = -\dfrac{\sqrt{a^2 - y^2}\,dy}{y}, \quad$ and $\quad x = -\displaystyle\int \frac{\sqrt{a^2 - y^2}\,dy}{y}$

The integration will give us a relation between x and y, which is the equation of the curve.

To effect the integration, let $u^2 = a^2 - y^2$, then

$$2u\,du = -2y\,dy, \quad \text{or} \quad u\,du = -y\,dy$$

$$x = \int \frac{u^2\,du}{y^2} = \int \frac{u^2\,du}{a^2 - u^2} = \int \frac{a^2 - (a^2 - u^2)}{a^2 - u^2} \cdot du$$

$$= a^2 \int \frac{du}{a^2 - u^2} - \int du$$

$$= a^2 \cdot \frac{1}{2a} \ln \frac{a + u}{a - u} - u + C$$

$$= \tfrac{1}{2}a \ln \frac{(a + u)(a + u)}{(a - u)(a + u)} - u + C$$

$$= a \ln \frac{a + u}{\sqrt{a^2 - u^2}} - u + C$$

We have then, finally,

$$x = a \ln \frac{a + \sqrt{a^2 - y^2}}{y} - \sqrt{a^2 - y^2} + C$$

When $x = 0$, $y = a$, so that $0 = a \ln 1 - 0 + C$, and $C = 0$; the equation of the tractrix is therefore

$$x = a \ln \frac{a + \sqrt{a^2 - y^2}}{y} - \sqrt{a^2 - y^2}$$

Example 5. Find the length of an arc of the logarithmic spiral $r = e^{\theta}$ between $\theta = 0$ and $\theta = 1$ radian.

Do you remember differentiating $y = e^x$? It is an easy one to remember, for it remains always the same whatever is done to it: $\frac{dy}{dx} = e^x$.

Here, since $\qquad r = e^{\theta}, \dfrac{dr}{d\theta} = e^{\theta} = r$

If we reverse the process and integrate $\int e^{\theta} \, d\theta$ we get back to $r + C$, the constant C being always introduced by such a process, as we have seen in **Chapter 17**.

It follows that

$$s = \int \sqrt{\left[r^2 + \left(\frac{dr}{d\theta} \right)^2 \right]} \, d\theta = \int \sqrt{(r^2 + r^2)} \, d\theta$$

$$= \sqrt{2} \int r \, d\theta = \sqrt{2} \int e^{\theta} \, d\theta = \sqrt{2} \, (e^{\theta} + C)$$

Integrating between the two given values $\theta = 0$ and $\theta = 1$, we get

$$s = \int_0^1 \sqrt{\left[r^2 + \left(\frac{dr}{d\theta} \right)^2 \right]} \, d\theta = \left[\sqrt{2}(e^{\theta} + C) \right]_0^1$$

$$= \sqrt{2} e^1 - \sqrt{2} e^0 = \sqrt{2}(e - 1)$$

$$= 1.41 \times 1.718 = 2.43$$

Example 6. Find the length of an arc of the logarithmic spiral $r = e^\theta$ between $\theta = 0$ and $\theta = \theta_1$.

As we have just seen,

$$s = \sqrt{2} \int_0^{\theta_1} e^\theta \, d\theta = \sqrt{2}[e^{\theta_1} - e^0] = \sqrt{2}(e^{\theta_1} - 1)$$

Example 7. As a last example let us work fully a case leading to a typical integration which will be found useful for several of the exercises found at the end of this chapter. Let us find the expression for the length of an arc of the curve $y = \dfrac{a}{2}x^2 + 3$.

$$\frac{dy}{dx} = ax, \quad s = \int \sqrt{1 + a^2x^2} \, dx$$

To work out this integral, let $ax = \sinh z$, then $a \, dx = \cosh z \, dz$, and $1 + a^2x^2 = 1 + \sinh^2 z = \cosh^2 z$;

$$s = \frac{1}{a} \int \cosh^2 z \, dz = \frac{1}{4a} \int (e^{2z} + 2 + e^{-2z}) dz$$

$$= \frac{1}{4a}\left[\tfrac{1}{2}e^{2z} + 2z - \tfrac{1}{2}e^{-2z}\right] = \frac{1}{8a}[(e^z)^2 - (e^{-z})^2 + 4z]$$

$$= \frac{1}{8a}(e^z - e^{-z})(e^z + e^{-z}) + \frac{z}{2a}$$

$$= \frac{1}{2a}(\sinh z \cosh z + z) = \frac{1}{2a}\left(ax\sqrt{1 + a^2x^2} + z\right)$$

To turn z back into terms of x, we have

$$ax = \sinh z = \tfrac{1}{2}(e^z - e^{-z})$$

Multiply out by $2e^z$,

$$2axe^z = e^{2z} - 1$$

or

$$(e^z)^2 - 2ax(e^z) - 1 = 0$$

This is a quadratic equation in e^z, and taking the positive root:

$$e^z = \tfrac{1}{2}\left(2ax + \sqrt{4a^2x^2 + 4}\right) = ax + \sqrt{1 + a^2x^2}$$

Taking natural logarithms:

$$z = \ln\left(ax + \sqrt{1 + a^2x^2}\right)$$

Hence, the integral becomes finally:

$$s = \int \sqrt{1 + a^2x^2}\,dx = \frac{x}{2}\sqrt{1 + a^2x^2} + \frac{1}{2a}\ln\left(ax + \sqrt{1 + a^2x^2}\right)$$

From several of the foregoing examples, some very important integrals and relations have been worked out. As these are of great use in solving many other problems, it will be an advantage to collect them here for future reference.

Inverse Hyperbolic Functions

If $x = \sinh z$, z is written inversely as $\sinh^{-1}x$;

and $\qquad z = \sinh^{-1}x = \ln\left(x + \sqrt{x^2 + 1}\right).$

Similarly, if $\quad x = \cosh z,$

$$z = \cosh^{-1}x = \ln\left(x + \sqrt{x^2 - 1}\right)$$

Irrational Quadratic Integrals

(i) $\displaystyle \int \frac{\sqrt{a^2 - x^2}}{x}\,dx = \sqrt{a^2 - x^2} - a\ln\frac{a + \sqrt{a^2 - x^2}}{x} + C$

(ii) $\displaystyle \int \sqrt{a^2 + x^2}\,dx = \tfrac{1}{2}x\sqrt{a^2 + x^2} + \tfrac{1}{2}a^2\ln\left(x + \sqrt{a^2 + x^2}\right) + C$

To these may be added:

(iii) $\displaystyle \int \frac{dx}{\sqrt{a^2 + x^2}} = \ln\left(x + \sqrt{a^2 + x^2}\right) + C$

For, if $x = a \sinh u$, $dx = a \cosh u \, du$, and

$$\int \frac{dx}{\sqrt{a^2 + x^2}} = \int du = u + C' = \sinh^{-1} \frac{x}{a} + C'$$

$$= \ln \frac{x + \sqrt{a^2 + x^2}}{a} + C'$$

$$= \ln \left(x + \sqrt{a^2 + x^2} \right) + C$$

You ought now to be able to attempt with success the following exercises. You will find it interesting as well as instructive to plot the curves and verify your results by measurement where possible.

Additional Problem Solving Examples

 The arc of the curve: y = ln x, lying in the fourth quadrant is revolved about the y-axis. Find the area of the surface generated.

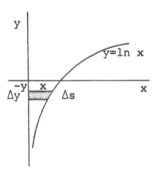

The shaded strip in the figure (x by Δy), when rotated about the y-axis, sweeps a volume approximately equal to that of a truncated cone, base radius x and thickness Δy, with a slanted edge Δs. If we were interested in finding the volume we would merely use Δy. However, for the surface area, Δs is involved, which converges to:
$$(ds)^2 = (dx)^2 + (dy)^2,$$
or,

$$ds = \left(1 = \left(\frac{dx}{dy}\right)^2\right)^{\frac{1}{2}} dy.$$

The surface area of an elementary disk, under the limit, is:

$$ds = 2\pi x \ ds$$

$$= 2\pi x \left(1 + \left(\frac{dx}{dy}\right)^2\right)^{\frac{1}{2}} dy.$$

$$y = \ln x \cdot \frac{dy}{dx} = \frac{1}{x} \cdot \frac{dx}{dy} = x.$$

$$dy = \frac{1}{x} \ dx.$$

Upon substitution,

$$ds = 2\pi x \ (1 + x)^2)^{\frac{1}{2}} \frac{1}{x} \ dx.$$

Integration from $x = 0$ to $x = 1$ yields the required area in the region bounded by the x- and y-axes.

$$s = 2\pi \int_0^1 \sqrt{1 + x^2} \ dx$$

$$= \pi \ \left(x\sqrt{1 + x^2} + \ln (x + \sqrt{1 + x^2})\right) \Big|_0^1$$

$$= \pi \ \left(\sqrt{2} + \ln (\sqrt{2} + 1)\right).$$

(Integration was carried out by the use of integration by parts and then substitution.)

 Find the length of the curve: $y = x^2$, from $x = 2$ to $x = 5$.

Applying the expression:

$$s = \int_a^b \sqrt{1 + \left(\frac{dy}{dx}\right)^2}\ dx,$$

we have $\quad y = x^2 \cdot y' = 2x.$

Therefore, $\quad s = \int_2^5 \sqrt{1 + 4x^2}\ dx.$

Integrating by parts and recalling

$$\int u\,dv = uv - \int v\,du,$$

let $\quad u = (1 + 4x^2)^{\frac{1}{2}}, \quad du = 4x(1 + 4x^2)^{-\frac{1}{2}},$

$\quad dv = dx, \quad v = x.$

$$\int_2^5 (1 + 4x^2)^{\frac{1}{2}}\ dx$$

$$= (1 + 4x^2)^{\frac{1}{2}} x - \int_2^5 \frac{4x^2}{(1 + 4x^2)^{\frac{1}{2}}}\ dx. \tag{1}$$

But $\quad \int_2^5 (1 + 4x^2)^{\frac{1}{2}}\ dx = \int_2^5 \frac{(1 + 4x^2)}{(1 + 4x^2)^{\frac{1}{2}}}\ dx.$

After expanding the right side, we have:

$$\int_2^5 (1 + 4x^2)^{\frac{1}{2}}\ dx$$

$$= \int_2^5 \frac{dx}{(1 + 4x^2)^{\frac{1}{2}}} + \int_2^5 \frac{4x^2 dx}{(1 + 4x^2)^{\frac{1}{2}}}. \tag{2}$$

By adding the equations (1) and (2), the term:

$$\int \frac{4x^2dx}{(1+4x^2)^{\frac{1}{2}}},$$

is eliminated.

At this point we obtain:

$$2 \int_{2}^{5}(1+4x^2)^{\frac{1}{2}} \, dx$$

$$= x(1+4x^2)^{\frac{1}{2}} + \int_{2}^{5} \frac{dx}{(1+4x^2)^{\frac{1}{2}}},$$

or, $\int_{2}^{5}(1+4x^2)^{\frac{1}{2}} \, dx$

$$= \frac{x}{2}(1+4x^2)^{\frac{1}{2}} + \int_{2}^{5} \frac{dx}{\left(\frac{1}{4}+x^2\right)^{\frac{1}{2}}}.$$

Note that

$$\int \frac{dx}{\left(\frac{1}{4}+x^2\right)^{\frac{1}{2}}}$$

is of the form:

$$\int \frac{dx}{(a^2+x^2)^{\frac{1}{2}}} = \ln\left(x+\sqrt{a^2-x^2}\right)$$

Thus,

$$\int \frac{dx}{\left(\frac{1}{4}+x^2\right)^{\frac{1}{2}}} = \ln\left(x+\sqrt{\frac{1}{4}-x^2}\right).$$

$$\int_{2}^{5}(1+4x^2)^{\frac{1}{2}} \, dx$$

reduces to: $\left[\dfrac{x}{2}(1+4x^2)^{\frac{1}{2}} + \ln\left(x + \sqrt{\dfrac{1}{4} - x^2}\right)\right]_2^5$

This expression, evaluated at x = 5, is equal to 27.43. At x = 2, it is equal to 5.52.

$$\int_2^5 (1+4x^2)^{\frac{1}{2}}\, dx = 27.43 - 5.52$$

$$= 21.91 \text{ units of arc}$$

Exercises XXII

(See answers on page 309)

(1) Find the length of the line $y = 3x + 2$ between the two points for which $x = 1$ and $x = 4$.

(2) Find the length of the line $y = ax + b$ between the two points for which $x = -1$ and $x = a^2$.

(3) Find the length of the curve $y = \frac{2}{3}x^{\frac{3}{2}}$ between the two points for which $x = 0$ and $x = 1$.

(4) Find the length of the curve $y = x^2$ between the two points for which $x = 0$ and $x = 2$.

(5) Find the length of the curve $y = mx^2$ between the two points for which $x = 0$ and $x = \dfrac{1}{2m}$.

(6) Find the length of the curve $x = a \cos \theta$ and $y = a \sin \theta$ between $\theta = \theta_1$ and $\theta = \theta_2$.

(7) Find the length of the arc of the curve $r = a \sec \theta$ from $\theta = 0$ to an arbitrary point on the curve.

(8) Find the length of the arc of the curve $y^2 = 4ax$ between $x = 0$ and $x = a$.

(9) Find the length of the arc of the curve $y = x\left(\dfrac{x}{2} - 1\right)$ between $x = 0$ and $x = 4$.

(10) Find the length of the arc of the curve $y = e^x$ between $x = 0$ and $x = 1$.

(*Note.* This curve is in rectangular coordinates, and is not the same curve as the logarithmic spiral $r = e^\theta$ which is in polar coordinates. The two equations are similar, but the curves are quite different.)

(11) A curve is such that the coordinates of a point on it are $x = a(\theta - \sin\ \theta)$ and $y = a(1 - \cos\ \theta)$, θ being a certain angle which varies between 0 and 2π. Find the length of the curve. (It is called a *cycloid*.)

(12) Find the length of an arc of the curve $y = \ln \sec x$ between $x = 0$ and $x = \dfrac{\pi}{4}$ radians.

(13) Find the expression for the length of an arc of the curve $y^2 = \dfrac{x^3}{a}$.

(14) Find the length of the curve $y^2 = 8x^3$ between the two points for which $x = 1$ and $x = 2$.

(15) Find the length of the curve $y^{\frac{2}{3}} + x^{\frac{2}{3}} = a^{\frac{2}{3}}$ between $x = 0$ and $x = a$.

(16) Find the length of the curve $r = a(1 - \cos\ \theta)$ between $\theta = 0$ and $\theta = \pi$.

You have now been personally conducted over the frontiers into the enchanted land. And in order that you may have a handy reference to the principal results, the author, in bidding you farewell, begs to present you with a passport in the shape of a convenient collection of standard forms. In the middle column are set down a number of the functions which most commonly occur. The results of differentiating them are set down on the left; the results of integrating them are set down on the right. May you find them useful!

Epilogue and Apologue

It may be confidently assumed that when this book falls into the hands of the professional mathematicians, they will (if not too lazy) rise up as one person and damn it as being a thoroughly bad book. Of that there can be, from their point of view, no possible manner of doubt whatever. It commits several most grievous and deplorable errors.

First, it shows how ridiculously easy most of the operations of the calculus really are.

Secondly, it gives away so many trade secrets. By showing you that *what one fool can do, other fools can do also,* it lets you see that these mathematical swells, who pride themselves on having mastered such an awfully difficult subject as the calculus, have no such great reason to be puffed up. They like you to think how terribly difficult it is, and don't want that superstition to be rudely dissipated.

Thirdly, among the dreadful things they will say about "So Easy" is this: that there is an utter failure on the part of the author to demonstrate with rigid and satisfactory completeness the validity of sundry methods which he has presented in simple fashion, and has even *dared to use* in solving problems! But why should he not? You don't forbid the use of a watch to every person who does not know how to make one? You don't object to the musician playing on a violin that he has not himself constructed. You don't teach the rules of syntax to children until they have already become fluent in the *use* of speech. It would be equally absurd to require general rigid demonstrations to be expounded to beginners in the calculus.

One other thing will the professed mathematicians say about this thoroughly bad and vicious book: that the reason why it is *so easy* is because the author has left out all the things that are really difficult. And the ghastly fact about this accusation is that—*it is true!* That is, indeed, why the book has been written—written for the legion of innocents who have hitherto been deterred from acquiring the elements of the calculus by the stupid way in which its teaching is almost always presented. Any subject can be made repulsive by presenting it bristling with difficulties. The aim of this book is to enable beginners to learn its language, to acquire familiarity with its endearing simplicities, and to grasp its powerful methods of solving problems, without being compelled to toil through the intricate out-of-the-way (and mostly irrelevant) mathematical gymnastics so dear to the unpractical mathematician.

There are amongst young engineers a number on whose ears the adage that *what one fool can do, another can,* may fall with a familiar sound. They are earnestly requested not to give the author away, nor to tell the mathematicians what a fool he really is.

Table Of Standard Forms

$\dfrac{dy}{dx}$	y	$\displaystyle\int y\,dx$
	*Algebraic*	
1	x	$\frac{1}{2}x^2 + C$
0	a	$ax + C$
1	$x \pm a$	$\frac{1}{2}x^2 \pm ax + C$
a	ax	$\frac{1}{2}ax^2 + C$
$2x$	x^2	$\frac{1}{3}x^3 + C$
nx^{n-1}	x^n	$\dfrac{1}{n+1}x^{n+1} + C$
$-x^{-2}$	x^{-1}	$\ln x + C$
$\dfrac{du}{dx} \pm \dfrac{dv}{dx} \pm \dfrac{dw}{dx}$	$u \pm v \pm w$	$\displaystyle\int u\,dx \pm \int v\,dx \pm \int w\,dx$
$u\dfrac{dv}{dx} + v\dfrac{du}{dx}$	uv	Put $v = \dfrac{dy}{dx}$ and integrate by parts
$\dfrac{v\dfrac{du}{dx} - u\dfrac{dv}{dx}}{v^2}$	$\dfrac{u}{v}$	No general form known
$\dfrac{du}{dx}$	u	$\displaystyle\int u\,dx = ux - \int x\,du + C$

Exponential and Logarithmic

e^x	e^x	$e^x + C$
x^{-1}	$\ln x$	$x(\ln x - 1) + C$
$0.4343x^{-1}$	$\log_{10}x$	$0.4343x(\ln x - 1) + C$
$a^x\ln a$	a^x	$\dfrac{a^x}{\ln a} + C$

Trigonometric

$\cos x$	$\sin x$	$-\cos x + C$
$-\sin x$	$\cos x$	$\sin x + C$
$\sec^2 x$	$\tan x$	$-\ln\cos x + C$

Circular (Inverse)

$\dfrac{1}{\sqrt{1-x^2}}$	arcsin x	$x \arcsin x + \sqrt{1-x^2} + C$
$-\dfrac{1}{\sqrt{1-x^2}}$	arccos x	$x \arccos x - \sqrt{1-x^2} + C$
$\dfrac{1}{1+x^2}$	arctan x	$x \arctan x - \frac{1}{2} \ln(1+x^2) + C$

Hyperbolic

$\cosh x$	$\sinh x$	$\cosh x + C$
$\sinh x$	$\cosh x$	$\sinh x + C$
$\operatorname{sech}^2 x$	$\tanh x$	$\ln \cosh x + C$

Miscellaneous

$-\dfrac{1}{(x+a)^2}$	$\dfrac{1}{x+a}$	$\ln|x+a| + C$
$-\dfrac{x}{(a^2+x^2)^{\frac{3}{2}}}$	$\dfrac{1}{\sqrt{a^2+x^2}}$	$\ln\left(x + \sqrt{a^2+x^2}\right) + C$
$\pm\dfrac{b}{(a \pm bx)^2}$	$\dfrac{1}{a \pm bx}$	$\pm\dfrac{1}{b} \ln|a \pm bx| + C$
$\dfrac{-3a^2 x}{(a^2+x^2)^{\frac{5}{2}}}$	$\dfrac{a^2}{(a^2+x^2)^{\frac{3}{2}}}$	$\dfrac{x}{\sqrt{a^2+x^2}} + C$
$a \cos ax$	$\sin ax$	$-\dfrac{1}{a} \cos ax + C$
$-a \sin ax$	$\cos ax$	$\dfrac{1}{a} \sin ax + C$
$a \sec^2 ax$	$\tan ax$	$-\dfrac{1}{a} \ln|\cos ax| + C$
$\sin 2x$	$\sin^2 x$	$\dfrac{x}{2} - \dfrac{\sin 2x}{4} + C$
$-\sin 2x$	$\cos^2 x$	$\dfrac{x}{2} + \dfrac{\sin 2x}{4} + C$

Miscellaneous

$n \cdot \sin^{n-1}x \cdot \cos x$	$\sin^n x$	$-\dfrac{\cos x}{n}\sin^{n-1}x$ $+\dfrac{n-1}{n}\displaystyle\int \sin^{n-2}x \, dx + C$
$-\dfrac{\cos x}{\sin^2 x}$	$\dfrac{1}{\sin x}$	$\ln\left|\tan\dfrac{x}{2}\right| + C$
$-\dfrac{\sin 2x}{\sin^4 x}$	$\dfrac{1}{\sin^2 x}$	$-\cot x + C$
$\dfrac{\sin^2 x - \cos^2 x}{\sin^2 x \cdot \cos^2 x}$	$\dfrac{1}{\sin x \cdot \cos x}$	$\ln|\tan x| + C$
$n \cdot \sin mx \cdot \cos nx +$ $m \cdot \sin nx \cdot \cos mx$	$\sin mx \cdot \sin nx$	$\dfrac{\sin (m-n)x}{2(m-n)} -$ $\dfrac{\sin (m+n)x}{2\,(m+n)} + C$
$a \sin 2ax$	$\sin^2 ax$	$\dfrac{x}{2} - \dfrac{\sin 2ax}{4a} + C$
$-a \sin 2ax$	$\cos^2 ax$	$\dfrac{x}{2} + \dfrac{\sin 2ax}{4a} + C$

Answers to Exercises I

(1) $\dfrac{dy}{dx} = 13x^{12}$ (2) $\dfrac{dy}{dx} = -\dfrac{3}{2}x^{-\frac{5}{2}}$ (3) $\dfrac{dy}{dx} = 2ax^{2a-1}$

(4) $\dfrac{du}{dt} = 2.4t^{1.4}$ (5) $\dfrac{dz}{du} = \dfrac{1}{3}u^{-\frac{2}{3}}$ (6) $\dfrac{dy}{dx} = -\dfrac{5}{3}x^{-\frac{8}{3}}$

(7) $\dfrac{du}{dx} = -\dfrac{8}{5}x^{-\frac{13}{5}}$ (8) $\dfrac{dy}{dx} = 2ax^{a-1}$

(9) $\dfrac{dy}{dx} = \dfrac{3}{q}x^{\frac{3-q}{q}}$ (10) $\dfrac{dy}{dx} = -\dfrac{m}{n}x^{-\frac{m+n}{n}}$

Answers to Exercises II

(1) $\dfrac{dy}{dx} = 3ax^2$ (2) $\dfrac{dy}{dx} = 13 \times \dfrac{3}{2}x^{\frac{1}{2}}$ (3) $\dfrac{dy}{dx} = 6x^{-\frac{1}{2}}$

(4) $\dfrac{dy}{dx} = \dfrac{1}{2}c^{\frac{1}{2}}x^{-\frac{1}{2}}$ (5) $\dfrac{du}{dz} = \dfrac{an}{c}z^{n-1}$ (6) $\dfrac{dy}{dt} = 2.36t$

(7) $\dfrac{dl_t}{dt} = 0.000012 \times l_0$

(8) $\dfrac{dc}{dV} = abV^{b-1}$, 0.98, 3.00 and 7.46 candle power per volt respectively.

(9) $\dfrac{dn}{dD} = -\dfrac{1}{LD^2}\sqrt{\dfrac{gT}{\pi\sigma}}, \dfrac{dn}{dL} = -\dfrac{1}{DL^2}\sqrt{\dfrac{gT}{\pi\sigma}}$

$\dfrac{dn}{d\sigma} = -\dfrac{1}{2DL}\sqrt{\dfrac{gT}{\pi\sigma^3}}, \dfrac{dn}{dT} = \dfrac{1}{2DL}\sqrt{\dfrac{g}{\pi\sigma T}}$

(10) $\dfrac{\text{Rate of change of } P \text{ when } t \text{ varies}}{\text{Rate of change of } P \text{ when } D \text{ varies}} = -\dfrac{D}{t}$

(11) 2π, $2\pi r$, πl, $\frac{2}{3}\pi r h$, $8\pi r$, $4\pi r^2$.

Answers to Exercises III

(1) (a) $1 + x + \dfrac{x^2}{2} + \dfrac{x^3}{6} + \dfrac{x^4}{24} + \ldots$ (b) $2ax + b$ (c) $2x + 2a$

(d) $3x^2 + 6ax + 3a^2$

(2) $\dfrac{dw}{dt} = a - bt$ (3) $\dfrac{dy}{dx} = 2x$

(4) $14110x^4 - 65404x^3 - 2244x^2 + 8192x + 1379$

(5) $\dfrac{dx}{dy} = 2y + 8$ (6) $185.9022654x^2 + 154.36334$

(7) $\dfrac{-5}{(3x + 2)^2}$ (8) $\dfrac{6x^4 + 6x^3 + 9x^2}{(1 + x + 2x^2)^2}$

(9) $\dfrac{ad - bc}{(cx + d)^2}$ (10) $\dfrac{anx^{-n-1} + bnx^{n-1} + 2nx^{-1}}{(x^{-n} + b)^2}$

(11) $b + 2ct$

(12) $R_0(a + 2bt), R_0\left(a + \dfrac{b}{2\sqrt{t}}\right), -\dfrac{R_0(a + 2bt)}{(1 + at + bt^2)^2}$ or $-\dfrac{R^2(a + 2bt)}{R_0{}^2}$

(13) $1.4340(0.000014t - 0.001024), -0.00117, -0.00107, -0.00097$

(14) (a) $\dfrac{dE}{dl} = b + \dfrac{k}{i}$, (b) $\dfrac{dE}{di} = -\dfrac{c + kl}{i^2}$

Answers to Exercises IV

(1) $17 + 24x$; 24 (2) $\dfrac{x^2 + 2ax - a}{(x + a)^2}$; $\dfrac{2a(a + 1)}{(x + a)^3}$

(3) $1 + x + \dfrac{x^2}{1 \times 2} + \dfrac{x^3}{1 \times 2 \times 3}$; $1 + x + \dfrac{x^2}{1 \times 2}$

(4) *(Exercises III):*

(1) (a) $\dfrac{d^2u}{dx^2} = \dfrac{d^3u}{dx^3} = 1 + x + \frac{1}{2}x^2 + \frac{1}{6}x^3 + \ldots$

     (b) $2a$, 0 (c) 2, 0 (d) $6x + 6a$, 6

(2) $-b$, 0 (3) 2, 0

(4) $56440x^3 - 196212x^2 - 4488x + 8192.$
     $169320x^2 - 392424x - 4488$

(5) 2, 0 (6) $371.80453x$, 371.80453

(7) $\dfrac{30}{(3x+2)^3}$, $\quad -\dfrac{270}{(3x+2)^4}$

(Examples):

(1) $\dfrac{6a}{b^2}x$, $\dfrac{6a}{b^2}$ (2) $\dfrac{3a\sqrt{b}}{2\sqrt{x}} - \dfrac{6b\sqrt[3]{a}}{x^3}$, $\dfrac{18b\sqrt[3]{a}}{x^4} - \dfrac{3a\sqrt{b}}{4\sqrt{x^3}}$

(3) $\dfrac{2}{\sqrt[3]{\theta^8}} - \dfrac{1.056}{\sqrt[5]{\theta^{11}}}$, $\dfrac{2.3232}{\sqrt[5]{\theta^{16}}} - \dfrac{16}{3\sqrt[3]{\theta^{11}}}$

(4) $810t^4 - 648t^3 + 479.52t^2 - 139.968t + 26.64$
     $3240t^3 - 1944t^2 + 959.04t - 139.968$

(5) $12x + 2$, 12 (6) $6x^2 - 9x$, $12x - 9$

(7) $\dfrac{3}{4}\left(\dfrac{1}{\sqrt{\theta}} + \dfrac{1}{\sqrt{\theta^5}}\right) + \dfrac{1}{4}\left(\dfrac{15}{\sqrt{\theta^7}} - \dfrac{1}{\sqrt{\theta^3}}\right)$

     $\cdot \dfrac{3}{8}\left(\dfrac{1}{\sqrt{\theta^5}} - \dfrac{1}{\sqrt{\theta^3}}\right) - \dfrac{15}{8}\left(\dfrac{7}{\sqrt{\theta^9}} + \dfrac{1}{\sqrt{\theta^7}}\right)$

Answers to Exercises V

(2) 64; 147.2; and 0.32 feet per second

(3) $\dot{x} = a - gt$; $\ddot{x} = -g$ (4) 45.1 feet per second

(5) 12.4 feet per second per second. Yes.

(6) Angular velocity $= 11.2$ radians per second;
     angular acceleration $= 9.6$ radians per second per second.

(7) $v = 20.4t^2 - 10.8, \quad a = 40.8t \quad 172.8 \text{ in./sec.}, 122.4 \text{ in./sec.}^2$

(8) $v = \dfrac{1}{30\sqrt[3]{(t-125)^2}}, a = -\dfrac{1}{45\sqrt[3]{(t-125)^5}}$

(9) $v = 0.8 - \dfrac{8t}{(4+t^2)^2}, a = \dfrac{24t^2 - 32}{(4+t^2)^3}, 0.7926 \text{ and } 0.00211$

(10) $n = 2, n = 11$

Answers to Exercises VI

(1) $\dfrac{x}{\sqrt{x^2+1}}$

(2) $\dfrac{x}{\sqrt{x^2+a^2}}$

(3) $-\dfrac{1}{2\sqrt{(a+x)^3}}$

(4) $\dfrac{ax}{\sqrt{(a-x^2)^3}}$

(5) $\dfrac{2a^2 - x^2}{x^3\sqrt{x^2-a^2}}$

(6) $\dfrac{\frac{3}{2}x^2\left[\frac{8}{9}x(x^3+a) - (x^4+a)\right]}{(x^4+a)^{\frac{2}{3}}(x^3+a)^{\frac{3}{2}}}$

(7) $\dfrac{2a(x-a)}{(x+a)^3}$

(8) $\frac{5}{2}y^3$

(9) $\dfrac{1}{(1-\theta)\sqrt{1-\theta^2}}$

Answers to Exercises VII

(1) $\dfrac{dw}{dx} = -\dfrac{3x^2(3+3x^3)}{27\left(\frac{1}{2}x^3 + \frac{1}{4}x^6\right)^3}$

(2) $\dfrac{dv}{dx} = -\dfrac{12x}{\sqrt{1+\sqrt{2+3x^2}}\left(\sqrt{3} + 4\sqrt{1+\sqrt{2+3x^2}}\right)^2}$

(3) $\dfrac{du}{dx} = -\dfrac{x^2\left(\sqrt{3}+x^3\right)}{\sqrt{\left[1+\left(1+\dfrac{x^3}{\sqrt{3}}\right)^2\right]^3}}$

(5) $\dfrac{dx}{d\theta} = a(1 - \cos \theta) = 2a \sin^2 \tfrac{1}{2}\theta$

$\dfrac{dy}{d\theta} = a \sin \theta = 2a \sin \tfrac{1}{2}\theta \cos \tfrac{1}{2}\theta$; $\dfrac{dy}{dx} = \cot \tfrac{1}{2}\theta$

(6) $\dfrac{dx}{d\theta} = -3a \cos^2 \theta \sin \theta$, $\dfrac{dy}{d\theta} = 3a \sin^2 \theta \cos \theta$;

$\dfrac{dy}{dx} = -\tan \theta$

(7) $\dfrac{dy}{dx} = 2x \cot (x^2 - a^2)$

(8) Write $y = u - x$; find $\dfrac{dx}{du}, \dfrac{dy}{du}$, and then $\dfrac{dy}{dx}$

Answers to Exercises VIII

(2) 1.44

(4) $\dfrac{dy}{dx} = 3x^2 + 3$; and the numerical values are: 3, $3\tfrac{3}{4}$, 6, and 15.

(5) $\pm\sqrt{2}$

(6) $\dfrac{dy}{dx} = -\dfrac{4}{9}\dfrac{x}{y}$. Slope is zero where $x = 0$; and is $\pm\dfrac{1}{3\sqrt{2}}$ where $x = 1$.

(7) $m = 4$, $n = -3$

(8) Intersections at $x = 1$, $x = -3$. Angles $153° \ 26'$, $2° \ 28'$

(9) Intersection at $x = y = \tfrac{25}{7}$. Angle $16° \ 16'$

(10) $x = \tfrac{1}{3}$, $y = 2\tfrac{1}{3}$, $b = -\tfrac{5}{3}$

Answers to Exercises IX

(1) Min.: $x = 0$, $y = 0$; max.: $x = -2$, $y = -4$

(2) $x = a$ (4) $25\sqrt{3}$ square inches.

(5) $\dfrac{dy}{dx} = -\dfrac{10}{x^2} + \dfrac{10}{(8-x)^2}$; $x = 4$; $y = 5$

(6) Max. for $x = -1$; min. for $x = 1$.

(7) Join the middle points of the four sides.

(8) $r = \frac{2}{3}R$, $r = \dfrac{R}{2}$, no max.

(9) $r = R\sqrt{\dfrac{2}{3}}$, $r = \dfrac{R}{\sqrt{2}}$, $r = 0.8507R$

(10) At the rate of $8\sqrt{\pi}$ square feet per second.

(11) $r = \dfrac{R\sqrt{8}}{3}$

Answers to Exercises X

(1) Max.: $x = -2.19$, $y = 24.19$; min.: $x = 1.52$, $y = -1.38$

(2) $\dfrac{dy}{dx} = \dfrac{b}{a} - 2cx$; $\dfrac{d^2y}{dx^2} = -2c$; $x = \dfrac{b}{2ac}$ (a *maximum*)

(3) (*a*) One maximum and two minima.
   (*b*) One maximum.   ($x = 0$; other points unreal.)

(4) Min.: $x = 1.71$, $y = 6.13$ (5) Max.: $x = -.5$, $y = 4$

(6) Max.: $x = 1.414$, $y = 1.7678$. Min.: $x = -1.414$,
   $y = -1.7678$

(7) Max.: $x = -3.565$, $y = 2.12$. Min.: $x = +3.565$, $y = 7.88$

(8) $0.4N$, $0.6N$ (9) $x = \sqrt{\dfrac{a}{c}}$

(10) Speed 8.66 nautical miles per hour. Time taken 115.44 hours, total cost is $2,251.11.

(11) Max. and min. for $x = 7.5$, $y = \pm 5.413$.

(12) Min.: $x = \frac{1}{2}$, $y = 0.25$; max.: $x = -\frac{1}{3}$, $y = 1.407$

Answers to Exercises XI

(1) $\dfrac{2}{x-3}+\dfrac{1}{x+4}$ (2) $\dfrac{1}{x-1}+\dfrac{2}{x-2}$ (3) $\dfrac{2}{x-3}+\dfrac{1}{x+4}$

(4) $\dfrac{5}{x-4}-\dfrac{4}{x-3}$ (5) $\dfrac{19}{13(2x+3)}-\dfrac{22}{13(3x-2)}$

(6) $\dfrac{2}{x-2}+\dfrac{4}{x-3}-\dfrac{5}{x-4}$

(7) $\dfrac{1}{6(x-1)}+\dfrac{11}{15(x+2)}+\dfrac{1}{10(x-3)}$

(8) $\dfrac{7}{9(3x+1)}+\dfrac{71}{63(3x-2)}-\dfrac{5}{7(2x+1)}$

(9) $\dfrac{1}{3(x-1)}+\dfrac{2x+1}{3(x^2+x+1)}$

(10) $x+\dfrac{2}{3(x+1)}+\dfrac{1-2x}{3(x^2-x+1)}$

(11) $\dfrac{3}{x+1}+\dfrac{2x+1}{x^2+x+1}$ (12) $\dfrac{1}{x-1}-\dfrac{1}{x-2}+\dfrac{2}{(x-2)^2}$

(13) $\dfrac{1}{4(x-1)}-\dfrac{1}{4(x+1)}+\dfrac{1}{2(x+1)^2}$

(14) $\dfrac{4}{9(x-1)}-\dfrac{4}{9(x+2)}-\dfrac{1}{3(x+2)^2}$

(15) $\dfrac{1}{x+2}-\dfrac{x-1}{x^2+x+1}-\dfrac{1}{(x^2+x+1)^2}$

(16) $\dfrac{5}{x+4}-\dfrac{32}{(x+4)^2}+\dfrac{36}{(x+4)^3}$

(17) $\dfrac{7}{9(3x-2)^2}+\dfrac{55}{9(3x-2)^3}+\dfrac{73}{9(3x-2)^4}$

(18) $\dfrac{1}{6(x-2)}+\dfrac{1}{3(x-2)^2}-\dfrac{x}{6(x^2+2x+4)}$

Answers to Exercises XII

(1) $ab(e^{ax} + e^{-ax})$ (2) $2at + \dfrac{2}{t}$ (3) $\ln n$ (5) npv^{n-1}

(6) $\dfrac{n}{x}$ (7) $\dfrac{3e^{-\frac{x}{x-1}}}{(x-1)^2}$ (8) $6xe^{-5x} - 5(3x^2 + 1)e^{-5x}$

(9) $\dfrac{ax^{a-1}}{x^a + a}$ (10) $\dfrac{15x^2 + 12x\sqrt{x} - 1}{2\sqrt{x}}$ (11) $\dfrac{1 - \ln(x+3)}{(x+3)^2}$

(12) $a^x(ax^{a-1} + x^a \ln a)$ (14) Min.: $y = 0.7$ for $x = 0.693$

(15) $\dfrac{1+x}{x}$ (16) $\dfrac{3}{x}(\ln ax)^2$

Answers to Exercises XIII

(2) $T = 34.625$; 159.45 minutes

(5) (*a*) $x^x(1 + \ln x)$;   (*b*) $2x\,(e^x)^x$;   (*c*) $e^{x^x} \times x^x\,(1 + \ln x)$

(6) 0.14 second   (7) (*a*) 1.642;   (*b*) 15.58

(8) $\mu = 0.00037$, 31.06 min

(9) i is 63.4% of i_0, 221.56 kilometers

(10) $k = 0.1339, 0.1445, 0.1555$, mean $= 0.1446$; percentage errors:—10.2%, 0.18% nil, $+71.8\%$.

(11) Min. for $x = \dfrac{1}{e}$ (12) Max. for $x = e$

(13) Min. for $x = \ln a$

Answers to Exercises XIV

(1) (i) $\dfrac{dy}{d\theta} = A \cos\left(\theta - \dfrac{\pi}{2}\right)$

(ii) $\dfrac{dy}{d\theta} = 2 \sin\theta \cos\theta = \sin 2\theta$ and $\dfrac{dy}{d\theta} = 2 \cos 2\theta$

303

(iii) $\dfrac{dy}{d\theta} = 3 \sin^2\theta \cos\theta$ and $\dfrac{dy}{d\theta} = 3 \cos 3\theta$

(2) $\theta = 45°$ or $\dfrac{\pi}{4}$ radians

(3) $\dfrac{dy}{dt} = -n \sin 2\pi nt$

(4) $a^x \, 1_n \, a \cos a^x$

(5) $\dfrac{-\sin x}{\cos x} = -\tan x$

(6) $18.2 \cos (x + 26°)$

(7) The slope is $\dfrac{dy}{d\theta} = 100 \cos (\theta - 15°)$, which is a maximum when $(\theta - 15°) = 0$, or $\theta = 15°$; the value of the slope being then $= 100$. When $\theta = 75°$ the slope is $100 \cos (75° - 15°) = 100 \cos 60° = 100 \times \frac{1}{2} = 50$

(8) $\cos\theta \sin 2\theta + 2 \cos 2\theta \sin\theta = 2 \sin\theta (\cos^2\theta + \cos 2\theta)$
$$= 2 \sin\theta (3 \cos^2\theta - 1)$$

(9) $amn\theta^{n-1} \tan^{m-1} (\theta^n) \sec^2 (\theta^n)$

(10) $e^x (\sin^2 x + \sin 2x)$

(11) (i) $\dfrac{dy}{dx} = \dfrac{ab}{(x+b)^2}$ (ii) $\dfrac{a}{b} e^{-\frac{x}{b}}$ (iii) $\dfrac{1}{90°} \times \dfrac{ab}{(b^2 + x^2)}$

(12) (i) $\dfrac{dy}{dx} = \sec x \tan x$ (ii) $\dfrac{dy}{dx} = -\dfrac{1}{\sqrt{1-x^2}}$

(iii) $\dfrac{dy}{dx} = \dfrac{1}{1+x^2}$ (iv) $\dfrac{dy}{dx} = \dfrac{1}{x\sqrt{x^2-1}}$

(v) $\dfrac{dy}{dx} = \dfrac{\sqrt{3} \sec x (3 \sec^2 x - 1)}{2}$

(13) $\dfrac{dy}{d\theta} = 4.6 (2\theta + 3)^{1.3} \cos (2\theta + 3)^{2.3}$

(14) $\dfrac{dy}{d\theta} = 3\theta^2 + 3\cos(\theta + 3) - \ln 3(\cos\theta \times 3^{\sin\theta} + 3^\theta)$

(15) $\theta = \cot\theta$; $\theta = \pm 0.86$; $y = \pm 0.56$; is max. for $+\theta$, min. for $-\theta$

Answers to Exercises XV

(1) $x^2 - 6x^2y - 2y^2$; $\frac{1}{3} - 2x^3 - 4xy$

(2) $2xyz + y^2z + z^2y + 2xy^2z^2$
$2xyz + x^2z + xz^2 + 2x^2yz^2$
$2xyz + x^2y + xy^2 + 2x^2y^2z$

(3) $\dfrac{1}{r}\{(x-a) + (y-b) + (z-c)\} = \dfrac{(x+y+z)-(a+b+c)}{r}$; $\dfrac{2}{r}$

(4) $dy = v\,u^{v-1}\,du + u^v \ln u\,dv$

(5) $dy = 3\sin v\,u^2\,du + u^3\cos v\,dv$
$dy = u\,(\sin x)^{u-1}\cos x\,dx + (\sin x)^u \ln\sin x\,du$
$dy = \dfrac{1}{v}\dfrac{1}{u}\,du - \ln u\,\dfrac{1}{v^2}\,dv$

(7) There is no minimum or maximum.

(8) (*a*) Length 2 feet, width = depth = 1 foot, vol. = 2 cubic feet

(*b*) Radius $= \dfrac{2}{\pi}$ feet = 7.64 in., length = 2 feet, vol. = 2.55

(9) All three parts equal; the product is maximum.

(10) Minimum $= e^2$ for $x = y = 1$

(11) Min. $= 2.307$ for $x = \frac{1}{2}$, $y = 2$

(12) Angle at apex $= 90°$; equal sides = length $= \sqrt[3]{2V}$

Answers to Exercises XVI

(1) $1\frac{1}{3}$.

(2) 0.6345

(3) 0.2624

(4) $y = \frac{1}{8}x^2 + C$

(5) $y = x^2 + 3x + C$

305

Answers to Exercises XVII

(1) $\dfrac{4\sqrt{a}\,x^{\frac{3}{2}}}{3}+C$

(2) $-\dfrac{1}{x^3}+C$

(3) $\dfrac{x^4}{4a}+C$

(4) $\frac{1}{3}x^3+ax+C$

(5) $-2x^{-\frac{5}{2}}+C$

(6) $x^4+x^3+x^2+x+C$

(7) $\dfrac{ax^2}{4}+\dfrac{bx^3}{9}+\dfrac{cx^4}{16}+C$

(8) $\dfrac{x^2+a}{x+a}=x-a+\dfrac{a^2+a}{x+a}$ by division. Therefore the answer

is $\frac{1}{2}x^2-ax+a(a+1)\ln(x+a)+C$

(9) $\dfrac{x^4}{4}+3x^3+\dfrac{27}{2}x^2+27x+C$

(10) $\dfrac{x^3}{3}+\dfrac{2-a}{2}x^2-2ax+C$

(11) $a^2\left(2x^{\frac{3}{2}}+\frac{9}{4}x^{\frac{4}{3}}\right)+C$

(12) $-\frac{1}{3}\cos\theta-\frac{1}{6}\theta+C$

(13) $\frac{1}{2}\theta+\dfrac{\sin 2a\theta}{4a}+C$

(14) $\frac{1}{2}\theta-\frac{1}{4}\sin 2\theta+C$

(15) $\frac{1}{2}\theta-\dfrac{\sin 2a\theta}{4a}+C$

(16) $\frac{1}{3}e^{3x}+C$

(17) $\ln|1+x|+C$

(18) $-\ln|1-x|+C$

Answers to Exercises XVIII

(1) Area $=120$; mean ordinate $=20$.

(2) Area $=\frac{4}{3}a^{\frac{5}{2}}$

(3) Area $=2$; mean ordinate $=2/\pi=0.637$

(4) Area $=1.57$; mean ordinate $=0.5$

(5) $0.571, 0.0476$

(6) Volume $=\frac{1}{3}\pi r^2 h$

(7) 1.25

(8) 79.6

(9) Volume $=4.935$, from 0 to π

(10) $a\ln a,\ \dfrac{a}{a-1}\ln a$

(12) A.M. $= 9.5$; Q.M. $= 10.85$

(13) Quadratic mean $= \dfrac{1}{\sqrt{2}} \sqrt{A_1{}^2 + A_3{}^2}$;

arithmetical mean $= 0$.

The first involves the integral

$$\int (A_1{}^2 \sin^2 x + 2A_1 A_3 \sin x \sin 3x + A_3{}^2 \sin^2 3x) dx$$

which may be evaluated by putting $\sin^2 x = \tfrac{1}{2}(1 - \cos 2x)$, $\sin^2 3x = \tfrac{1}{2}(1 - \cos 6x)$ and $2 \sin x \sin 3x = \cos 2x - \cos 4x$.

(14) Area is 62.6 square units. Mean ordinate is 10.43.

(16) 436.3 (This solid is pear-shaped.)

Answers to Exercises XIX

(1) $\dfrac{x\sqrt{a^2 - x^2}}{2} + \dfrac{a^2}{2} \arcsin \dfrac{x}{a} + C$

(2) $\dfrac{x^2}{2} \left(\ln x - \tfrac{1}{2} \right) + C$

(3) $\dfrac{x^{a+1}}{a+1} \left(\ln x - \dfrac{1}{a+1} \right) + C$

(4) $\sin e^x + C$

(5) $\sin (\ln x) + C$

(6) $e^x(x^2 - 2x + 2) + C$

(7) $\dfrac{1}{a+1} (\ln x)^{a+1} + C$

(8) $\ln |\ln x| + C$

(9) $2 \ln |x - 1| + 3 \ln |x + 2| + C$

(10) $\tfrac{1}{2} \ln |x - 1| + \tfrac{1}{5} \ln |x - 2| + \tfrac{3}{10} \ln |x + 3| + C$

(11) $\dfrac{b}{2a} \ln \left| \dfrac{x - a}{x + a} \right| + C$

(12) $\ln \left| \dfrac{x^2 - 1}{x^2 + 1} \right| + C$

(13) $\tfrac{1}{4} \ln \left| \dfrac{1 + x}{1 - x} \right| + \tfrac{1}{2} \arctan x + C$

(14) $-\dfrac{\sqrt{a^2 - b^2 x^2}}{b^2} + C$

Answers to Exercises XX

(1) $T = T_0 e^{\mu\theta}$ $\qquad\qquad$ (2) $s = ut + \frac{1}{2}at^2$

(3) Multiplying out by e^{2t} gives $\dfrac{d}{dt}(ie^{2t}) = e^{2t}\sin 3t$, so that,

$$ie^{2t} = \int e^{2t}\sin 3t\, dt = \tfrac{1}{13}e^{2t}(2\sin 3t - 3\cos 3t) + E$$

Since $i = 0$ when $t = 0$, $E = \frac{3}{13}$; hence the solution becomes $i = \frac{1}{13}(2\sin 3t - 3\cos 3t + 3e^{-2t})$.

Answers to Exercises XXI

(1) $r = 2\sqrt{2}$, $x_1 = -2$, $y_1 = 3$ $\quad$ (2) $r = 2.83$, $x_1 = 0$, $y_1 = 2$

(3) $x = \pm 0.383$, $y = 0.147$ $\qquad$ (4) $r = \sqrt{2|m|}$, $x_1 = y_1 = 2\sqrt{m}$

(5) $r = 2a$, $x_1 = 2a$, $y_1 = 0$

(6) When $x = 0$, $r = y_1 = $ infinity, $x_1 = 0$
When $x = +0.9$, $r = 3.36$, $x_1 = -2.21$, $y_1 = +2.01$
When $x = -0.9$, $r = 3.36$, $x_1 = +2.21$, $y_1 = -2.01$

(7) When $x = 0$, $r = 1.41$, $x_1 = 1$, $y_1 = 3$
When $x = 1$, $r = 1.41$, $x_1 = 0$, $y_1 = 3$
Minimum $= 1.75$

(8) For $x = -2$, $r = 112.3$, $x_1 = 109.8$, $y_1 = -17.2$
For $x = 0$, $r = x_1 = y_1 = $ infinity
For $x = 1$, $r = 1.86$, $x_1 = -0.67$, $y_1 = -0.17$

(9) $x = -0.33$, $y = +1.07$

(10) $r = 1$, $x = 2$, $y = 0$ for all points. A circle.

(11) When $x = 0$, $r = 1.86$, $x_1 = 1.67$, $y_1 = 0.17$
When $x = 1.5$, $r = 0.365$, $x_1 = 1.59$, $y_1 = 0.98$
$x = 1$, $y = 1$ for zero curvature.

(12) When $\theta = \dfrac{\pi}{2}$, $r = 1$, $x_1 = \dfrac{\pi}{2}$, $y_1 = 0$.

When $\theta = \dfrac{\pi}{4}$, $r = 2.598$, $x_1 = 2.285$, $y_1 = -1.414$

(14) When $\theta = 0$, $r = 1$, $x_1 = 0$, $y_1 = 0$

When $\theta = \dfrac{\pi}{4}$, $r = 2.598$, $x_1 = -0.715$, $y_1 = -1.414$

When $\theta = \dfrac{\pi}{2}$, $r = x_1 = y_1 = $ infinity

(15) $r = \dfrac{(a^4 y^2 + b^4 x^2)^{\frac{3}{2}}}{a^4 b^4}$, when $x = 0$, $y = \pm b$, $r = \dfrac{a^2}{b}$,

$x_1 = 0$, $y_1 = \pm \dfrac{b^2 - a^2}{b}$; when $y = 0$, $x = \pm a$, $r = \dfrac{b^2}{a}$,

$x_1 = \pm \dfrac{a^2 - b^2}{a}$, $y_1 = 0$

(16) $r = 4a \ |\sin \tfrac{1}{2}\theta|$

Answers to Exercises XXII

(1) $s = 9.487$ (2) $s = (1 + a^2)^{\frac{3}{2}}$ (3) $s = 1.22$

(4) $s = \displaystyle\int_0^2 \sqrt{1 + 4x^2}\, dx$

$= \left[\dfrac{x}{2}\sqrt{1 + 4x^2} + \tfrac{1}{4}\ln\left(2x + \sqrt{1 + 4x^2}\right) \right]_0^2 = 4.65$

(5) $s = \dfrac{0.57}{m}$ (6) $s = a(\theta_2 - \theta_1)$ (7) $s = \sqrt{r^2 - a^2}$

(8) $s = \displaystyle\int_0^a \sqrt{1 + \dfrac{a}{x}}\, dx$ and $s = a\sqrt{2} + a \ln\left(1 + \sqrt{2}\right) = 2.30a$

(9) $s = \dfrac{x-1}{2}\sqrt{(x-1)^2+1} + \tfrac{1}{2}\ln\left\{(x-1)+\sqrt{(x-1)^2+1}\right\}$

and $s = 6.80$

(10) $s = \displaystyle\int_1^e \dfrac{\sqrt{1+y^2}}{y}\,dy.$ Put $u^2 = 1+y^2$; this leads to

$$s = \sqrt{1+y^2} + \ln\dfrac{y}{1+\sqrt{1+y^2}} \text{ and } s = 2.00$$

(11) $s = 4a\displaystyle\int_0^\pi \sin\dfrac{\theta}{2}\,d\theta$ and $s = 8a$

(12) $s = \displaystyle\int_0^{\frac{1}{4}\pi} \sec x\,dx.$ Put $u = \sin x$; this leads to

$s = \ln\left(1+\sqrt{2}\right) = 0.8814$

(13) $s = \dfrac{8a}{27}\left\{\left(1+\dfrac{9x}{4a}\right)^{\frac{3}{2}} - 1\right\}$

(14) $s = \displaystyle\int_1^2 \sqrt{1+18x}\,dx.$ Let $1+18x = z$, express s in terms

of z and integrate between the values of z corresponding to $x = 1$ and $x = 2$. $s = 5.27$

(15) $s = \dfrac{3a}{2}$ (16) $4a$

All earnest students are exhorted to manufacture more examples for themselves at every stage, so as to test their powers. When integrating they can always test their answer by differentiating it, to see whether they get back the expression from which they started.

Definite Integrals

1. $\displaystyle\int_0^\infty x^n e^{-ax} dx = \frac{n!}{a^{n+1}}$

2. $\displaystyle\int_1^\infty \frac{dx}{x^a} = \frac{1}{a-1}$, $a > 1$

3. $\displaystyle\int_0^\infty \frac{dx}{(1+x)x^a} = \pi\csc a\pi$, $0 < a < 1$

4. $\displaystyle\int_0^\infty \frac{x^{a-1}}{1+x} dx = \frac{\pi}{\sin a\pi}$, $0 < a < 1$

5. $\displaystyle\int_0^\infty \frac{a\,dx}{a^2+x^2} = \begin{cases} \dfrac{\pi}{2} & \text{if } a > 0 \\[2mm] 0 & \text{if } a = 0 \\[2mm] -\dfrac{\pi}{2} & \text{if } a < 0 \end{cases}$

6. $\displaystyle\int_0^{\pi/2} \cos^n x \, dx = \int_0^{\pi/2} \sin^n x \, dx = \begin{cases} \dfrac{1\cdot 3\cdot 5\cdots(n-1)}{2\cdot 4\cdot 6\cdots n}\ \dfrac{\pi}{2} & \text{for } n \text{ even} \\[3mm] \dfrac{2\cdot 4\cdot 6\cdots(n-1)}{1\cdot 3\cdot 5\cdots n} & \text{for } n \text{ odd} \end{cases}$

7. $\displaystyle\int_0^\infty \frac{\sin x}{x} \, dx = \frac{\pi}{2}$

8. $\displaystyle\int_0^\infty \frac{\cos ax}{1+x^2} \, dx = \frac{\pi}{2} e^{-|a|}$

311

9. $\displaystyle\int_0^\pi \sin nx \sin mx \; dx = \int_0^\pi \cos nx \cos mx \, dx = 0$

   m,n are integers and m≠n

10. $\displaystyle\int_0^\pi \sin^2 nx\, dx = \int_0^\pi \cos^2 nx \; dx = \frac{\pi}{2}$

11. $\displaystyle\int_0^\infty \frac{\sin^2 x}{x} \; dx = \frac{\pi}{2}$

12. $\displaystyle\int_0^\infty \sin^2 x \; dx = \int_0^\infty \cos^2 x \; dx = \frac{1}{2}\left(\frac{\pi}{2}\right)^{\frac{1}{2}}$

13. $\displaystyle\int_0^\infty \frac{\sin x}{x^{\frac{1}{2}}} \; dx = \int_0^\infty \frac{\cos x}{x^{\frac{1}{2}}} \; dx = \left(\frac{\pi}{2}\right)^{\frac{1}{2}}$

14. $\displaystyle\int_0^\infty e^{-ax} \; dx = \frac{1}{a} \quad , \quad a > 0$

15. $\displaystyle\int_0^\infty \frac{e^{-ax}-e^{-bx}}{x} \; dx = \ln \frac{b}{a} \qquad a,b > 0$

16. $\displaystyle\int_0^\infty e^{-a^2 x^2} dx = \frac{1}{2a} \; \pi^{\frac{1}{2}}$

17. $\displaystyle\int_0^\infty x e^{-x^2} \; dx = \frac{1}{2}$

18. $\displaystyle\int_0^\infty x^2 e^{-x^2} dx = \frac{\pi^{\frac{1}{2}}}{4}$

19. $\int_0^\infty x^2 e^{-ax^2} dx = \frac{\pi^{\frac{1}{2}}}{4} a^{-3/2}$

20. $\int_0^\infty e^{-a^2x^2} \cos x \, dx = \frac{\pi}{2|a|} \, e^{-1/4 \, a^2}, \, a \neq 0$

21. $\int_0^\infty x[\mathrm{erf}(ax)] e^{-b^2x^2} dx = \frac{a}{2b^2} (a^2+b^2)^{-\frac{1}{2}}$

22. $\int_0^\infty x^2[\mathrm{erf}(ax)] e^{-b^2x^2} dx = \frac{1}{2\pi^{\frac{1}{2}}} \left[\frac{a}{b^2(a^2+b^2)} + \frac{1}{b^3} \tan^{-1}\frac{a^2}{b} \right]$

23. $\int_0^\infty [\mathrm{erf}(ax)] e^{-b^2x^2} dx = \frac{1}{(\pi b^2)^{\frac{1}{2}}} \tan^{-1}\frac{a}{b}$

24. $\int_0^\infty [1-\mathrm{erf}(ax)] dx = \frac{1}{a \, \pi^{\frac{1}{2}}}$

25. $\int_0^1 (\ln x)^n dx = (-1)^n n!$

26. $\int_0^1 \frac{\ln x}{1+x} dx = \frac{\pi^2}{12}$

27. $\int_0^1 \frac{\ln x}{(1-x^2)^{\frac{1}{2}}} dx = -\frac{\pi}{2} \ln 2$

28. $\int_0^{\pi/2} \ln(\tan x) dx = 0$

29. $\int_0^\pi x \ln(\sin x) dx = -\frac{\pi^2}{2} \ln 2$

313

INDEX

317

trigonometric functions, 163
trigonometric integrals, 243
triumphs, 237

U

undetermined constant, 191, 196–197

V

variable, 11
  dependent, 15
  independent, 15
    time as, 55–63
varying, 20
velocity, 56, 60–63
volume, *xii,* 219

W

work, 58

What has been said about this book

(Continued from front two pages)

" This is the very best math book I have ever read,
An excellent source for understanding calculus! "
Teacher from Washington, D.C.

" I wish I had been introduced to this book when I took
math in my first year of college. "
Student from Sacramento, California

" A truly well-written book. "
Student from Smithfield, North Carolina

" This book has the power to teach anybody simple calculus. I
heartily recommend this book because of its simplicity in
explanation, and clearness of direction. "
A Calculus reader

" An excellent piece on a most difficult topic. "
Student from San Luis Obispo, California

" Make sure you put your name on your copy of this book.
Everyone else will want to borrow it from you. "
Student from Bayonne, New Jersey

" Excellent for Beginners. "
Student from St. Louis, Missouri

(More on next page)

What has been said about this book

(Continued from previous page)

" This book explains the philosophy of the subject in a very simple manner, making it easy to understand even for people who are not proficient in math. "

Teacher from Bolivia

" The book introduces calculus at the right pace and in a way that makes it very interesting. "

Student from Paris, France

" If you want to understand calculus, get this book. "

Student from San Diego, California

" I really can't stress the fact enough that this math book is easy to understand. "

Student from Stewartsville, NJ

" The book still inspires me to dig deeper into the calculus. The writing is succinct and the illustrations are quite good. "

Student from San Diego, California

" The book is fantastic, but you must be mathematically eager and know some algebra. Many other books on calculus go too far, and are too broad for the reader to get the basics in a concise way. "

Student, Edinburgh, United Kingdom

(More on front two pages)